TOMATOES

TOMATOES
Terry Marshall

Whittet Books

West Dean Gardens

The cover photograph was one of several varieties photographed at West Dean Gardens, West Dean, Chichester, West Sussex PO18 OQZ. This wonderful garden, open to the public, has 150 varietes of tomato, growing, on display and also hosts a tomato day.
tel no: 01243 818210
website: www.westdean.org.uk .

Acknowledgments

My thanks are due to all those who have in any way contributed to the preparation of this book. A special thank you to Annabel Whittet for her guiding hand and for her photographs of varieties at West Dean Gardens.

I am indebted to Mr. Philip Waite, Head Gardener at Wimpole Hall NT, for his help and permission to photograph his range of varieties. To Mr. Neil Porteous, Head Gardener at Clumber Park NT., for his help and selection of cultivars.

My thanks to the HDRA Seed Library for maintaining such a wide range of the heritage varieties. To Horticulture Research International Wellesbourne for so many research reports. To the John Innes Institute, of Norwich for permission to quote the work of Mr. W.J. C. Lawrence.

A very special thank you to my wife Jennifer for all her help and support in growing, tasting, and evaluating many hundreds of tomato varieties over the years.

Thanks to the Organic Gardening Catalogue for their help with pictures.

Thanks to Thompson and Morgan for permission to use the picture of Ailsa Craig (p.87) and Green Grape (p.122) and Association Kokopelli for their picture of Black Crimea (p.115).

CONTENTS

Introduction

Enjoyed for decades as a seasonal treat, tomatoes are now widely available every week of the year. Fast transport systems, often covering hundreds of miles, ensure a continuous supply of fruit to shops and supermarkets regardless of season. Today small varieties are presented for sale in transparent pre-weighed packs, standard sizes are sold by weight and large beefsteaks individually.

The range of varieties is steadily increasing especially among the cherry types, which look colourful, bright and inviting. Shelves full of attractive tomatoes are the result of the plant breeders' search for the perfect commercial cultivar, they are a testimony to diligent scientific research and the devotion and skill of the grower.

But what about flavour, what do they taste like? How does buying and eating a supermarket tomato compare with picking and eating a succulent, sun ripened 'old' variety or a modern cultivar straight from the plant? Are modern F1 varieties as good as their ancestors and with taste being so very subjective what exactly constitutes a good tomato variety and where do you look to find your personal favourite?

In this book I have tried to include as wide a selection of tomato types, sizes, colours and potential flavours as possible; some are well over a century old, some new for 2006, but all are available as seeds or plants. To select just 100 varieties from the thousands that are available worldwide is a difficult choice to make, but it comes back to those two words 'potential flavour'.

Tomato flavour depends on several factors and when well grown all the varieties listed are capable of developing a flavour in keeping with their type, size and colour. Among the qualities found in this selection are large fruit with thin skins and an abundance of luscious juice; black cultivars with a spicy, smoky, exotic 'tang' and flavourful yet mild, low acid, and white varieties for those with a delicate palate.

Some are capable of withstanding the ravages of disease; some are very tasty; others are tough and will keep and travel well when off the plant. The very names of some of the old varieties tantalize the taste buds with memories of tomatoes picked and eaten in the family greenhouse. A new range of cherry tomatoes has been developed that is ideally suited to include in the lunch box and is attractive to children in size, shape and flavour.

Whatever the qualities of commercial varieties and however swiftly they are distributed, they are unlikely to match the flavour and pleasure of eating your own warm, sweet, juicy fruit. Select a variety that appeals to you, choose a way of growing it to suit your facilities then grow and enjoy this most delicious of fruits. There is no other plant that we can grow that offers as much vitamin and mineral rich fruit, over such a long season as the tomato. Here are some truly wonderful varieties to grow and savour.

1 Origins

The tomato as we know it is descended from the wild species found in the northern arm of the Andes in South America, an area including Ecuador, Colombia, Peru, Bolivia and Chile. The sprawling vine *Lycopersicon pimpinellifolium* with tiny sweet fruit still grows wild at the western end of the river valleys in Peru but is seldom found at high altitudes. It seems that the earlier Andean peoples did not cultivate or use the small fruited vines for there is no word for tomato in the South American Indian languages, nor as yet have any drawings or pottery markings of tomatoes been found from that era.

Over centuries the wild strains were carried northwards into central America and Mexico, in all probability by travellers, birds and animals. It was the people of Meso-America who originally worked on improving the wild strains by selecting the best fruiting plants. At this time nature took a hand and spontaneous mutation occurred. The basic materials for tomato genetics are mutant genes and the earliest available variants were found as varietal characteristics such as different sizes, shapes and colours. By selecting and planting the seeds of larger and better tasting fruit, a whole range of strains were developed.

Mexico's Aztecs and Mayans cultivated 'Xitomatle' which later became 'Tomati' or 'Tomatle'. In the UK we say 'Tomato', to the French, Germans and Spanish it is 'Tomate', the Dutch add another 'a' to make it 'Tomaat', while the American say 'Tom-A-to'. However it is pronounced, the root goes back to 'Tomatl' from the Nahatal language of the Aztec people.

Exactly when the tomato arrived in Europe is still unknown. The office of The Director of the Archivo General de Indias in Seville, Spain, says that the actual arrival has not been definitely documented and cannot be given a specific date. What is well documented is that between 1519 and 1521 the Spanish army, led by Hernan Cortes, invaded and colonised Mexico and, as was their practice, collected seeds as they went.

The Conquistadors as they were known found that the Aztecs were cultivating a yellow-fruited mutant tomato which they prized above all the other strains that they were growing. The tomato seed carried across the Atlantic in those stately Spanish galleons during the 1520s-1530s was likely to have come from quite small fruit of a yellowish-orange colour. Successive voyages returned with a motley collection of seeds from fruit of different sizes shapes and colours. Many of the sub-tropical strains did not survive the European weather, nor did they breed true, with some strains bitter and hard.

The Spaniards introduced the new fruit to Italy when Naples came under Spanish rule in 1522, and it must have been the golden form that arrived in Naples for in 1544 Pier Andrea Mattioli wrote in his herbal about the introduction of *Mala-aurea* – golden apples 'Which resemble a Melrose apple, flattened and segmented, green at first and when ripe of a golden colour which is eaten in the same manner as an egg plant – fried in oil with salt and pepper'. Ten years later in a second edition to his herbal, Mattioli tells of the introduction of a red variety of the fruit and notes that the Italian name for the Latin *Mala aurea* is *Pomidoro,* which is still used in Italy today.

However, people were becoming very wary of the new fruit. Botanists identified the plant as a member of the Solanaceae family which includes Henbane, Deadly Nightshade and Mandrake and there was a suspicion that the fruit might be as poisonous as its dreaded cousins.

2 Classification

Tomato plants arrived in England around 1550 when they were first grown as ornamental curiosities by a few aristocrats in their houses. The first known British tomatogrower was Patrick Bellow in 1554. In 1574, writing in his herbal, Rembert Dodens of Antwerp claimed 'The tomatoes are eaten by some, prepared and cooked with salt pepper and oil. They offer the body very little nourishment and that unwholesome'. Dodens' conclusion appears to have been accepted without question by herbalists for many years to come. John Gerard (1545-1612), who was for 20 years consultant gardener to that great Elizabethan statesman Lord Burghley, planted the first culinary tomatoes in England while superintendant of the College of Physicians Garden in Holborn in the 1590s. *Gerard's Herbal* was published in 1597 and, while he knew they were edible, Gerard repeated Dodens' conclusion and added, 'They are ranke, stinking and poisonous,' an opinion which was often repeated over the next two centuries. With the tomato regarded as inedible in Britain and northern Europe, nevertheless the stems and leaves were used by herbalists to treat skin conditions.

The herbalist John Parkinson (1567-1650), an apothecary to King James I, published his book *Theatrum Botanicum* in 1640, which was for many years the most comprehensive English book on medicinal plants. Parkinson describes the two kinds of tomato known in his day as, 'One, the "Great Apple of Love" bearing fruite which are of the bigness of a small or mean pippin, of a faire pale reddish colour, like unto an orange, full of slimie juice and watery pulp'. He describes the other as 'The fruits are small, round, yellowish-red berries not much bigger than great grapes' these he called 'Small Love Apples'. Parkinson goes on to say, 'Wee onely have them for a curiosity in our gardens and for the amorous aspect or beauty of the fruite. In the hot countries where they naturally grow they are much eaten of the people to coole and quench the heate and thirst of their hot stomachs. The apples also, boyled or infused in oyle in the sun is thought to be good to cure the itch it will allay the heate thereof'. This is probably the first recorded definition of the now popular sun dried tomatoes in oil.

During the 16th century botanists gave the tomato the Latin name of *Lyco-persicon*, – 'Wolf peach' – because they thought that although it looked like a peach it was unfit to eat. At one time it was considered that the tomato belonged to the genus *Solanum*. In fact when Linnaeus introduced his Binomial Nomenclature of Plants in 1749 using specific adjectives, he classified the tomato within this genus as 'Solanum lycopersicon'. Nineteen years later Phillip Miller in his *Gardener's Directory* of 1768 changed the species name to *esculentum*, meaning edible. Although the subject is still debated, *Solanum* and *Lycopersicon* are now considered to belong to separate genus based on different leaf types and pollen dispersal. The full classification of the tomato is now accepted as:°

Order	Scrophulariales
Suborder	Solanineae
Family	*Solanaceae*
Tribe	*Solaneae*
Genus	*Lycopersicon*
Subgenus	*Eulycopersicon*
Species	*Lycopersicon esculentum*
Variety	Ailsa Craig

3 Tomato development

It was the Italians who did much of the early work of tomato breeding and development. The oldest botanical garden in the world was established in 1545 by the University of Padua to the West of Venice in northern Italy; here botanists were keen to select, breed and grow a whole range of plants that would thrive there.

Some five years later Daniel Barbaro of the university ordered a Viridarium or greenhouse to be built at Padua. These structures were built of stone, brick or wood and contained a fireplace or portable brazier to burn coal, wood or charcoal to provide heat. As early as 1490 Pontanus (1426-1503) mentions that coal was being burned in wooden sheds to over-winter citrus plants. The name 'Greenhouse' was coined to describe their function, that was, to house green plants, oranges, lemons, myrtles and tender plants during the winter months.

Over the next two centuries patient breeding and careful selection turned the open flower characteristics of some of the mutants into the flowers seen in today's open pollinated (self-fertile) tomatoes. In modern varieties the style with its pollen receptive stigma is completely enclosed within the anther cone in the middle of the flower.

With an enclosed anther cone the pollen from the flower drops on to the adjacent central receptive stigma and self-pollination is ensured. Self-pollination means that existing open pollinated varieties such as Ailsa Craig and Gardeners Delight are all true breeding inbred lines, with the same varietal characteristics coming true from one generation to the next, so seed saved from the plants will reproduce the same characteristics as their parents.

Italian botanists and scientists selected and bred strains with the distinctive characteristics that eventually became beefsteaks or oxhearts and the famous plum types which today fill the tins of their canning industry and provide the base for the tubes of tomato concentrate.

4 Light

Looking at the natural habitat of the wild species of tomatoes it is obvious that while some are found growing at latitudes of up to approximately 25° N of the equator, the largest concentration are found between 5° and 18° South of the meridian, in the western arm of the Andes. Many of the habitats are in areas of high light intensity so it is not surprising that the tomato is a light sensitive plant. The tomato is also a heat sensitive plant coming from areas such as Lima in Peru which has an average daily temperature from a maximum of 66°F/27°C in August to a maximum of 83°F/33°C during February and March with a variation between these two during the rest of the year.

When growing tomato plants as a crop, light and heat are interconnected, one cannot work effectively without the other. Light falling on the leaves starts the process of photosynthesis in which light energy is converted into chemical energy. The energy produced combines atmospheric carbon dioxide(CO_2) with the water in the leaf to produce carbohydrates and ultimately all the materials for plant growth.

Light is a limiting factor for tomato plant growth particularly in winter as the plants only start to photosynthesise when the light levels are above 200 foot candles in intensity. The basic unit of light quantity is called the 'Lumen'; it can be measured in lumens per square foot, foot candles (fc), (that is, the light produced on a vertical screen at a distance of one foot from a standard candle) or, lumens per square metre (lux).

In winter the actual intensity of light falling on the seedlings may be only in the 250-300 foot candle range. As gardeners our eyes are ever optimistic. Human eyes are sensitive to light in the 450-650nm(nano-meter) wavelength range, peaking at 550nm and so it is easy to think that the light in a greenhouse is brighter than it actually is at plant level. Tomato plants photosynthesise in the 400-700nm wavelength range peaking at 670, so it is not just the quantity of light that they respond to but also the quality. Try checking light levels in the open garden, then at greenhouse plant leaf level in January with a light meter. Some light meters are calibrated in foot candles, others in lux; most camera meters clearly indicate the light levels and all too often it is a very dismaying experience. To heat a greenhouse for so little light is clearly uneconomic and it is far better to delay sowing until the light levels in any particular glasshouse are high enough to sustain continuous growth.

Depending on the time of the year, the intensity of bright sunshine in Britain ranges from 1,000-8,000 foot candles of light, but when the sun is obscured by cloud the intensity of the diffused light is infinitely variable. A combination of sunlight and diffused light provides the natural daylight for plant growth, with at least 70% of winter days being illuminated by diffused light. During late December through January on average only the middle 5 hours of daylight may be strong enough to reach 500 foot candles.

In Britain the average hours of sunshine over the last 40 years illustrate the wide variation between the North and South and coastal and inland areas. On the South coast around Eastbourne the district can expect around 1,840 hours of sunshine a year, while Edinburgh can expect around 1,400 (some 434 hours less than the South coast). In coastal districts and areas near large expanses of water there is the added advantage of reflected light in the sky. The glasshouse growing area around Blackpool with 1,533 hours of sunshine a year is slightly better than the glasshouse area of East Yorkshire which has around 1,500 hours per annum. Wisley, the home of the Royal Horticultural Society Gardens, with a 40-year average of some 1,534 hours of sunshine a year fares better than Sutton Bonnington in the middle of the country which has around 1,388 sunny hours per annum. Districts not subject to fog or atmospheric pollution enjoy better light levels; polluted air not only restricts the passage of light but also stains greenhouse glass thus reducing its transparency.

This may all sound rather academic but it is essentially practical. When growing tomatoes anywhere in the British Isles or further afield, timing seed sowing and planting to make the best use of local light values is the key to good crops of any variety.

Light in greenhouses

Greenhouses offer plants protection. Just how effective that protection is in terms of light and heat depends on many factors, but, large or small, the same natural laws apply to them all. Hardly a day goes by without some reference being made in the media to global warming and the 'greenhouse effect', yet greenhouses are deliberately designed and built to create and maximise the use of the greenhouse effect. It works like this.

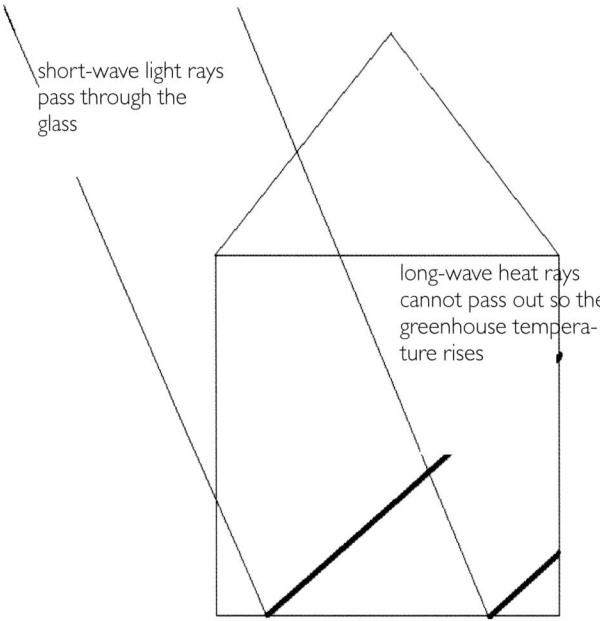

short-wave light rays pass through the glass

long-wave heat rays cannot pass out so the greenhouse tempera-ture rises

The greenhouse effect

Glass transmits a large part of the short-wave radiation (mostly visible light) from the sun, but is opaque to the infra-red radiation (heat rays) given out by bodies at temperatures of less than 100° Centigrade. The short-wave radiation from the sun passes through the greenhouse glass and is absorbed by, and warms up the soil, plants, paths, walls and internal structures inside the greenhouse; as the temperature of these is way below 100°C they give off heat rays which cannot pass out through the glass but are trapped inside and the air temperature rises higher than that of the outside air. The increase in temperature in an unheated glasshouse is roughly proportionate to the amount of light (direct and diffused) passing through the glass.

Light quality

It is the intensity, quality and duration of this combined amount of direct and diffused light actually penetrating into a cold greenhouse that determines when and what may be grown in it. For most of the year outdoor light values are appreciably higher than those found inside. A regular check with a light meter is not just revealing but positively dismaying. This is because light travels in straight lines. When a beam of light travels parallel to the surface of a sheet of glass none will pass through it. The percentage of light that will pass through a sheet of glass depends on the angle of the glass to the rays of light falling on it. This is called the angle of incidence, as seen in diagram 1.

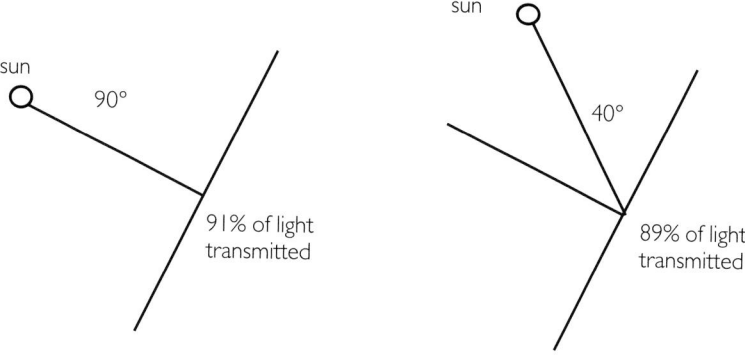

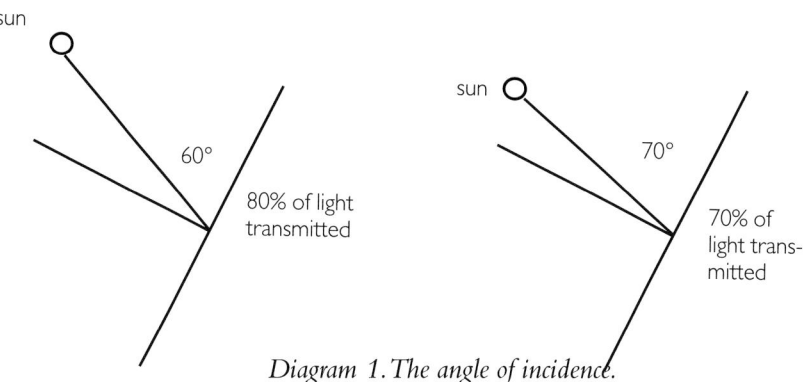

Diagram 1. The angle of incidence.

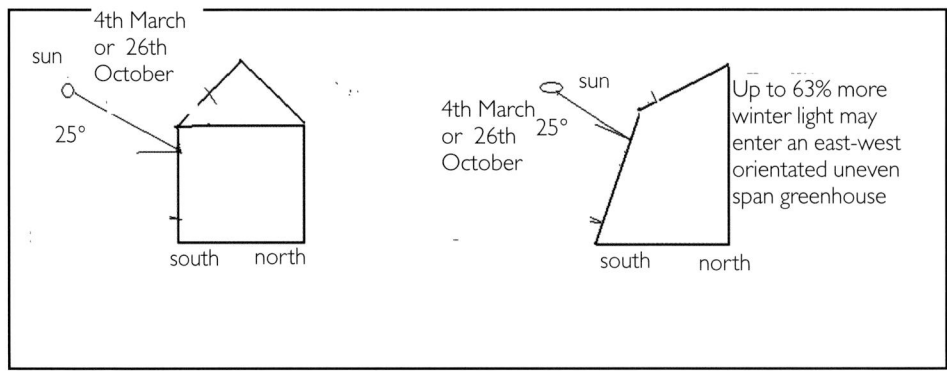

Diagram 2   Winter light (after W.J.C. Lawrence)

This angle of incidence depends on the design, orientation and construction of the glasshouse, as shown in diagram 2. Here the 60° sloping side contributes to improving the angle of incidence, allowing more of the low winter sunshine to pass through.

So light is the key. Without sufficient light strong enough to raise the ambient temperature of a cold greenhouse to growing point and intense enough to promote photosynthesis, plants, particularly tomato plants will not grow.

During the 7-8 weeks each side of the shortest day, the mean light of the sun above the unobstructed horizon is 12°, at the 52° line of latitude that crosses England from Felixstowe in Suffolk to Fishguard in Wales. North of this line the sun becomes progressively lower in the sky. So for winter propagation the angle of the sun to the glass is vital. Equally important is the actual orientation of the greenhouse.

Traditionally, glasshouses were built North to South to provide equal light to both sides of benches and beds. It was Mr W.J.C. Lawrence of the John Innes Horticultural Institute who in 1948 demonstrated that up to 27% more winter light could be gained: he went on to prove that when the glass on the South side of the house was built at 60°, the light transmission could be increased to by 63%.

Glasshouse design may well be more important in the future than it has ever been. For anyone thinking of building a new greenhouse for winter/spring natural daylight propagation, an uneven span East-West greenhouse is not only a more efficient light transmitter which is more economical to heat, but with the changing weather pattern creating more dull winter days may well become imperative. So it follows that when building a new greenhouse the more unobstructed light that the site has the better it will be and when the house is to be used for winter propagation an East-West orientation will gather more of the available light and produce better and quicker results.

Greenhouses are often inherited with the property and their site and structure cannot be moved, in this case it helps to keep them as free as possible from surrounding foliage.

New site or old, clean glass can make all the difference to the speed of growth and the health of greenhouse plants. It is a simple but very practical fact that dirty glass can restrict the transmission of light by up to 48%. In many parts of the country in the days of coal fires and factory chimneys, greenhouses wore a winter overcoat of soot, and glass cleaning was a weekly job. Today, in many districts traffic fumes create an insidious haze which falls imperceptibly on to glass and eventually build up into a shading layer, so regular winter cleaning may still be needed. A long handled very soft brush and soapy water works wonders. It is interesting to clean half a glasshouse then rinse it off and take a light meter reading in each half; the results are often startling.

A greenhouse provides protection for the plants that it contains, but only really works when the plants in it start to actively photosynthesise. Photosynthesis depends on the quantity and quality of the light transmitted through the glass; clean glass allows the process to start earlier in the day when light levels may be lower. Winter light levels are low and may fluctuate throughout the day; with clean glass most of the available light will be transmitted. The intensity of light affects the shape, size and thickness of a tomato leaf; low winter light produces small, delicate, paper thin leaves which are easily damaged making them liable to infection. A bench full of sturdy well developed tomato plants in early spring demonstrates the skill and ability of the propagator to balance the available light and heat to provide the plants with the best possible start.

Summary

•The importance of light cannot be overstressed.
•The speed and growth of a tomato plant is determined by the quality and quantity of light falling on the leaves.
•Over most of the country sunlight in Nov, Dec, Jan. is weak and sporadic with diffused light of low intensity.
•An E-W orientated greenhouse admits more winter light than a N–S orientated one.
•An uneven span E –W greenhouse captures the most winter light .
•Clean glass admits up to 48% more winter light than dirty glass.

.

5 Heating

Although light is the determining factor in photosynthesis, temperature is the other vital necessity for tomato plant growth and development. Given strong light and a temperature of 60° F/16°C plus, the rate of photosynthesis increases as the temperature rises. Creating a growing temperature in the greenhouse is a balance between the daily light levels and the facility to create artificial heat. The heating system chosen should be reliable and consistent during frosty days and nights, yet controllable and flexible when bright sunshine lifts the air temperature. Equipment fitted with thermostatic control can keep the temperature to within a degree or two of the desired figure.

Modern commercial tomato growers enjoy the benefits of computer controlled, high speed, fully flexible heating systems that are efficient and economical. For smaller growers and keen amateurs electricity is the simplest and most efficient form of heating; properly installed to comply with the safety regulations, it provides instant and flexible heat. Many growers, however, find that natural gas fired greenhouse heaters are the most economical way to heat their propagating houses. They do need to have a natural gas supply fitted to the site by an approved gas fitter, so the determining cost may be the distance from the main gas supply to the point of use. Where a mains supply is impractical, the widely available supply of bottled gas is an alternative. Compact free standing heaters are available with different types for mains or bottled gas, with models covering a range of heat outputs suitable for different sizes of greenhouses. Heaters are fitted with a pilot light and a thermostat which covers a selection of settings to provide warmth from frost protection up to propagating heat. Superb plants can be grown using gas heat, but heaters must be used carefully following the maker's ventilation instructions.

Gas by-products

The combustion by-products of burning gas are: carbon dioxide which helps plant growth, and water vapour which may cause condensation. Other phytotoxic by-products, nitric oxide (NO) and nitrogen dioxide (NO_2), form in the burner flame; the hotter it burns the more is produced. A spontaneous reaction with oxygen takes place in the greenhouse air and more nitrogen dioxide forms. At high concentrations the bottom leaves of the plant go brown, shrivel and die. If it continues the rest of the plant is affected. This is easily seen. What is not so obvious is the effect of medium concentrations of toxic gasses which reduce photosynthesis, slow growth and lower the yield. For all the effort put into the crop later in the season the damage has already been done and total crop weight suffers. Not all cultivars react in the same way: Ailsa Craig and Moneymaker are particularly sensitive to toxic gasses, while Sonato and some other Dutch varieties are less so. It seems to be a contradiction to spend money heating a greenhouse then leaving some ventilation open but it is one of those unquantifiable equations: sufficient vent to get rid of toxic by-products, but not enough to lose a lot of expensive heat in order to gain crop weight. It is very much a case of following the manufacturer's specific instructions for the particular model being used. Gas burners are constantly being improved to increase efficiency and reduce the harmful effects of combustion.

For short-term warmth, modern portable paraffin heaters are very efficient when low sulphur, premium grades of paraffin are burnt. A wide range of paraffin heaters is available to suit all sizes of small structures, but once again it is vital to carefully follow the maker's instructions regarding ventilation. Bubble plastic is often used to insulate propagating areas and can significantly reduce the cost of heating but it is all too easy to get too enthusiastic and create an airtight greenhouse. Tomato plants need several changes of air an hour, especially when the heat is from gas or paraffin heaters.

Summary

•Cost, consider, capital installation, and running costs of the alternative systems and fuels bearing in mind that fuel prices will continue to rise.
•It is essential to ensure that electrical and gas supplies are correctly installed in keeping with the latest regulations.
•Heating systems should have the capacity to cope with the coldest conditions.
•Thermostatic control will avoid under or over heating.
•Carefully follow the equipment maker's ventilation instructions.

6 Crop timing

Light strong enough for photosynthesis combined with heat, warm enough for plant growth, are the basic requirements for tomato growing. Then comes the question of timing and when is the most economical time to sow tomatoes in the British Isles.

In a fully heated glasshouse local light values are the determining factor. In warm, frost free, or cold glasshouses, to sow too early is a waste of time and money, while sowing too late is to lose valuable growing time in what is all too often a short growing season. In the past trial and error have resulted in arriving at local sowing dates often tied in to some specific day or family birthday, but the weather pattern is now changing.

All our accumulated horticultural expertise, glasshouse design and management to date has evolved during weather conditions reflecting our maritime climate. Over the centuries the effect of the Gulf Stream and fast-moving weather systems over the British Isles have determined the climate and created our temperate type of flora. The quickly moving pressure fronts meant that whatever the season extreme conditions rarely occurred. During recent years however there have been wide fluctuations in temperature values, with several periods of the tissue-desiccating heat usually found in a hot Continental climate. Deluges of unusually heavy rain have caused severe local flooding in unexpected parts of the country. Prolonged droughts have resulted in water shortage of both surface and ground water, reservoirs have dried up and aquifers become badly depleted.

A worrying aspect from a glasshouse growing point of view is the increasing frequency of periods of low light intensity at any time of the year caused by dense cloud cover. There is nothing new about gloomy skies, they are part of our climate, but long periods of low cloud were usually more frequent during the winter months when there were few plants growing under glass. When poor light continued into March and April it was called 'a poor spring'. Meteorological stations throughout the country record local sunshine hours, but few continuously monitor light levels. During recent years there have been 5/7/10 days in succession when the light intensity at plant leaf level has been too low for photosynthesis to take place. When the light levels alter significantly, tomato leaves adjust to the new status, then take a couple of days to revert back to their previous capacity to photosynthesise when the intensity of the light values return. These are 'lost' growing days. When the sum total of lost, irreplaceable days between February and May reach 30, then 25% of potential growing time has been lost for the season.

Is this a temporary climatic phase or is it the reality of global warming on these islands? However the eventual pattern develops, at present in many areas plants are growing in what are often longer periods of low light levels than in the past. If the weather improves, all well and good, but if not, it is with the present light values that plants will have to be grown. Crop timing has always been important but it may well become imperative if the present weather pattern becomes the established norm.

When a well propagated tomato plant is around 8 weeks old, the flowers on the first truss open, but unless the temperature is high enough for the flowers to develop sufficient viable pollen, fertilisation does not occur, and they will not produce fruit.

North or South, East or West, no matter where they are grown, without pollen there is no fruit. A truss of tomato flowers opens over several days; if none of them makes pollen then the entire truss is barren and the time, effort and money that has gone into growing that all important first early truss has been wasted.

In the British Isles the 10° of latitude between 50° N at the tip of Cornwall, to 60° N in South Shetland covers such a wide range of climatic conditions and altitudes that sowing times from South to North can vary by up to as much as 3 months. But wherever a glasshouse is sited and whatever the level of heating is available, tomato plants in any structure will have a pollination date and from that a sowing and planting date can be calculated.

Pollination date is when in any greenhouse, anywhere, it is reasonable to expect to have a temperature of around 65° F/17°C upwards for several hours in the day on several days of the week. When viable pollen is transferred from the anthers to the receptive stigma in the centre of the flower by shaking or spraying, fertilisation takes place and fruit production begins. Broadly speaking it takes around 12-14 weeks in winter to propagate a tomato plant from seed sowing to planting stage under reasonable natural light. In late winter to early spring this reduces to 8 weeks. The shortest propagating time is in late spring when 6-7 weeks after sowing, some of the finest plants of the year will be ready for their final quarters. Using this time scale and matching propagating facilities to a known or calculated likely pollination date, seed can be confidently sown as follows.

Crop	Pollination date	Sow
Very early	Early February	Early November
Early with early varieties	Mid-March	1st January
Main crop, early varieties	End March	14th January
Main crop, most varieties	Mid-April	1st February
Main crop for cool house	End April	14th February
Cold house	Early May	14th March
Outdoor, large plants	May-June	21st March

Summary

- Time sowing to pollination date.
- For early crops choose varieties that have been bred for early work.
- Check recent local winter and spring sunshine records.
- Record sowing dates, pollination dates and crop results.
- After first season timing can be adjusted if necessary.

7 Supplementary lighting

Sometimes a heated glasshouse is available to plant up with a crop of early tomatoes but the propagating facilities are unsuitable for raising early plants because of low January and early February light levels due to site or shading by buildings; in this case it is worth considering supplementary lighting.

The earliest tomato crops of all are grown from seed sown during November and December when the natural light intensity is so low that propagation by artificial light has become standard commercial practice. Commercial growers have used lights for many years and under carefully controlled conditions grow their plants to an exact date. After the initial propagating stage commercial production is often based on a 'Replacement Lighting' system which totally or partially replaces daylight with artificial light. This means that fully automated 'growing rooms' can be constructed in any suitable building where temperatures can be maintained to within a degree. Lamps with the right spectral quality and intensity are hung above the seedlings for 12-16 hours a day. After 21 days the seedlings will have grown into fine healthy plants.

When plants receive 'supplementary lighting' the aim is to 'top up' poor natural light, or to extend the day length to speed up the growth rate, or a combination of the two. For commercial installations, high pressure sodium discharge lamps offer a balanced spectral output from a compact luminaire which only casts a small shadow in daylight.

The smaller grower and keen amateur, who only needs to extend the season by some 4–5 weeks, may find that a combination of a mercury vapour lamp and fluorescent tubes are the answer to their short-term requirements. A 160 watt mercury vapour lamp bulb fitted in a reflector and suspended some 15"-18"(38-45cm) over 2 trays of tomato seedlings will cast a pool of light much stronger than January daylight and save 10-14 days of growing time.

For economy and simplicity Gro-lux fluorescent tubes are well worth considering; they emit light similar to daylight and for 14-21 days tomato plants thrive under them, growing from seedlings into sturdy young plants. The tube or tubes are fitted to a simple frame or grid and spaced according to the light level required. The frame is then suspended over the plants on chains and raised, a link at a time, as the plants grow. The main problem with tubular rigs is the shadow they cast during sunny days when the lights are switched off, but if their main use is from mid-January to mid-February, the number of sunny days in inland areas is often far too few. Full details of all the lighting systems are available from the Farm Energy Centre.

The speed at which plants grow under artificial light depends on the type and intensity of the light source and the temperature at plant level, which should be in the 68°-73° F(20°-23°C) range when the lights are on. Young tomato plants need 8 hours of darkness at night, so when 12-16 hours of light is being used, an automatic on-off clock may be needed.

Replacement or supplementary lighting reduces the length of time needed to heat larger areas for propagation, but when supplementary lighting is available, there is a temptation to sow too soon. As with natural daylight propagation, seed is sown to produce plants in flower for the earliest pollination date that the intended growing glasshouses allow. Artificial lighting simply speeds up the process of propagation by several weeks with the added bonus of good truss development in the very top of the plant.

To allow the plant to continue to develop after being 'lit up', it needs to be placed in the lightest position on the benches, allowing sufficient space between plants so that their leaves do not overlap. It takes a couple of days or so for the light receptors on the surface of the leaves to adjust to the hours and intensity of the new daylight regimen.

From then on the plants are treated the same as daylight raised plants, but by comparison they may appear to be softer and leafier before settling down in their new environment.

Summary

•Would the increase in crop weight be worth the capital and running costs involved?
•Is supplementary lighting needed to improve crops?
•Lighting rigs need to be safely installed.
•Beware of being too early for local daylight levels for post 'lit' plants.
•Is space available for early post 'lit' plants?

8 Germination

Tomato seeds vary in price from 1p each for some open pollinated varieties to over 50p per seed for the latest F1 introductions. Even the expensive seed can be good value for money in view of the potential crop they are capable of yielding over a long season. But when sowing an expensive variety the true price is the number of plants raised from any given packet of seed. Whatever the cost of the seed, a high rate of germination can be expected by providing the optimum conditions needed by tomato seeds to germinate.

The seed for any specific tomato crop is sown at a time to make maximum use of the available propagating and growing facilities. Propagation usually fits into one of five time bands.

November-January

When grown in natural daylight this is the trickiest, most expensive propagating period, but may also be the most rewarding with early fruit and a potentially heavy crop from plants with a long season of growth ahead.

February-March

With naturally improving light intensity, March 21st is the pivot, when the speed of growth increases and as the daytime temperatures start to rise better plants are produced while heating costs are reduced.

March-April

An increase in daylight from 12 to 15 hours provides ideal conditions to grow strong sturdy plants for cold greenhouse cropping. This is the time when within days benches become crowded and plants need spacing out.

April-May

The easiest propagating time of all, with good light and fast growing conditions for raising late cold house and outdoor plants. At this time of year seedlings pricked out 6 to a tray or box, or into 5 inch/12.5cm pots or large modules should be ready for planting a month later.

May-June

Time to sow early varieties for short late summer crops, or for second crops to follow a very early crop that has been stopped for lack of space. Successful propagation depends on temperature control and keeping the pots moist. Long days and high temperatures can result in tall plants with up to 10 leaves before the first truss forms, but with careful timing they can go on to yield 4–5 trusses of good quality autumn fruit.

Moisture

Whenever tomato seed is sown the same basic needs have to be met: warmth, water and air are required for successful germination. Straight from a sealed packet the seed usually has a moisture content of 5.5%; it takes several hours for the seed coat, or testa, to absorb sufficient water and become turgid. During this time the living seed needs free oxygen to be able to respire and start the process of growth; this begins when the turgid seed has been exposed to sufficient warmth to break dormancy. Once the dormant state is broken, the seed must not be allowed to dry out. If needed, gently spray the seed tray with tepid water. The water enters and swells the seed by the micropyle which is that little scar on the side of the seed from which the tap root or 'radicle' emerges and starts to grow downwards. The radicle is already growing down by the time the small white loop of the hypocotyl pushes through the surface of the compost. The hypocotyl, which is the first part of the stem to be seen, straightens out revealing the seed leaves or cotyledons, which start to expand, heralding the start of a new season or the next tomato crop.

Seed quantities

Just how much seed is sown owes more to economics than to applied botany. When a lot of plants are required, or during the fast propagating period of late spring, sowing thickly and pricking out the seedlings has much to recommend it. Tomato seed is sold by weight or seed count: 7,000 seeds to the ounce used to be the guide, but small cherry varieties are well above this figure, while some of the modern F1 seed weighs heavy. Once a variety has been chosen, suppliers will usually provide details of numbers and expected germination percentages from a given weight of seed. The standard seed tray 14"x 9" x 2"(355 x 228 x 50mm) will provide space for about 250 seedlings if they are to be pricked off as soon as they can be handled.

A clean seed tray is filled with moist seed compost and the surface lightly pressed to a level, the tray is well watered and allowed to drain; 270-300 seeds, which should produce at least 250 fine seedlings, are scattered evenly over the surface of the compost. Cover the seed by lightly riddling ¼"(½cm) of the same compost over the box, and gently spray with tepid water.

Germination

Once sown, the seed tray is covered with a sheet of glass or hard plastic or put in a plastic bag and placed wherever a constant temperature can be maintained. Different parts of a greenhouse bench usually have different temperatures and it is worth moving a maxi-mini thermometer around to check it out before sowing. Cover the tray to exclude light as most varieties of tomato germinate better in darkness. Thermostatically controlled propagating units, where the temperature can be constantly maintained to within a degree, simplify the process as well as reducing the need to keep the whole greenhouse at the higher temperature.

For many gardeners however the airing cupboard is the most economical and consistent heat source for their early sowings. Most cupboards are far too hot both day and night so move a thermometer around to check the heat, often the door has to be left open to reduce the temperature but a few days trial and error before sowing will determine the best place to put the seed tray. The more exact and consistent the temperature can be maintained the quicker and more even the germination. Heritage and older varieties may germinate over a period of 2-3 weeks, but some of the newer F1 hybrids will start to appear after 3-4 days so check morning and evening to prevent the newly emerged seedlings from becoming drawn and spindly. Once germination has taken place the seedlings should be placed in a good light position, but not initially into bright sunlight.

Germination temperatures

Tomatoes will germinate over a wide range of temperatures, but the optimum is between 65°F/18°C and 68°F/20°C). The closer they are kept to this level the quicker and more consistent the germination will be. When the temperature is 10°F degrees lower(58°F/15°C), the seedlings take longer to come through and the speed of germination is often erratic with emergence taking place over up to 3 weeks. Conversely when the temperature is up in the high 70s, particularly under low light conditions, some open pollinated cultivars will produce 'rogues'; although these are thought to be a genetic tendency, they appear to be triggered by too–high germination temperatures. 'Rogues' are easily spotted during propagation as the numerous leaves grow opposite each other. From above, the leaves on a rogue plant appear to form a square; from the side they resemble a miniature Christmas tree. The plants are sterile, waste time and valuable bench space and are better removed when spotted.

Pricking out

Thickly sown seedlings need to be pricked out just as soon as they are large enough to be carefully handled by their seed leaves. Tomatoes heave 2 seed leaves or cotyledons, which are usually extended by the 5th or 6th day after emergence. Seedlings pricked out as early as the 5th or 6th day after germination grow into much better plants than those left to become drawn and 'leggy' before being moved. Seedlings pricked out directly into 5"/13cm pots in springtime make the best plants. A clean pot is filled with moist compost and struck off level, a hole is made and a seedling held gently by the seed leaf is eased from the seed tray and lowered into the hole. The compost is then placed round the stem with the slightest of pressure and the pot tapped on the bench, this settles the level of the compost which should be about ½"/1cm from the top of the pot, with the seedling showing ¼"/5mm of stem above the compost level.

Time was when plants were left for 24 hours after potting up in moist soil 'To make the roots find the water', but research has shown that seedlings that were watered in immediately after potting grow into far better plants. A gentle overhead spray settles the compost in the pot and provides water through the cotyledons which replaces and reduces transpirational loss. Transplanting shock is minimised and the roots continue growing unchecked.

Space sowing

In winter when the growth rate is slow and heated bench space is usually at a premium, space sown or box grown seedlings are often an acceptable compromise between available facilities and total crop weight. A clean seed tray is filled with seed compost and lightly pressed to a level surface, it is soaked or stood in water then allowed to drain. As the water seeps out of the compost air is pulled into the pore spaces to replace the drained water; it is worth noting that seed sown in saturated compost and kept at high temperatures take much longer to germinate. 'Only sow plump seeds', the old gardeners used to say, and when only a few plants are required or there is plenty of seed to choose from, they do appear to produce better plants. Spread the seed on a dark background and select the thickest plumpest seeds with the brightest seed coats from those available. Indentations are made in the surface of the compost 1½-2"/3.5-5cm apart, some ¼"/6mm deep; place a seed in each dent and cover with the same compost, water lightly to settle them in. Sowing at ½"/12mm deep used to be practised as the extra depth of compost almost always dragged the seed coat from off the emerging cotyledons which did not then come into contact with any virus that may have been on it. With the introduction of tomato mosaic resistant varieties this is not as critical as it was in the past. The seed tray is then covered with a sheet of glass, or hard plastic or put inside a plastic bag and placed wherever a constant germinating temperature can be maintained.

Potting on

Once germinated, space sown seedlings are placed in the best possible light position in the greenhouse, where, enjoying their space, they will grow into fine sturdy young plants. When the leaf development reaches the point where the leaves start to touch overlap and shade one another it is time to transfer them to 5"/13cm pots (in which they can stay until planted in final positions). The day before potting soak the tray of young plants with a solution of liquid seaweed diluted to the maker's instructions. This acts as a bio-stimulant, each plant taking with it a root system charged with minerals, trace elements and rooting stimulants, this will reduce the stress of trans-planting and minimise the check to continuous growth. Not only space sown, but all tomato plants benefit from a feed of liquid seaweed the day before potting on. Take a clean 5"/13cm pot, fill it ⅓ full of moist potting compost, ease out a young plant from the seed tray, it will have filled its space with a mass of roots, so lift the root ball intact and place in the pot adding potting compost to within ¼"/0.5 cm of the seed leaves and ½"/1.5cm from the top of the pot. A well grown space sown young plant should still have its cotyledons – seed leaves, those first narrow, thin flat leaves – intact at the time of potting, their presence indicates that the plant has been stress free. Any tomato seedling exposed to sudden cold conditions, toxic fumes, or more likely water stress, when hot sunshine dries the compost out, reacts by discarding its cotyledons. Usually the leaf scar heals completely and no wound is left; if the cotyledons are missing at the time of potting up sink the plant lower in the pot then top up with compost allowing the plant's natural stem rooting ability to develop an extended root system above the existing one.

Direct seed sowing

When only several plants are needed, sowing 2 or 3 seeds directly into a 4¼" or 5"/13cm pot and pulling out the weakest seedlings means that the remaining strong specimen never suffers any root disturbance or potting up shock. The result is a first-class plant grown in the simplest way possible. The one problem with this method is the nutrient level of the potting compost; this should contain sufficient nutrient for the propagating period but is often too 'strong' for emerging seedlings and may actu-ally inhibit germination or even burn the tip of the radicle. Seed compost contains just sufficient nutrient at a very low level suited to the needs of the seedling, when warmth, moisture, air, and the physical structure of the compost are more important than nutrients over those first few days. The same applies to peat-free, but do follow the manufacturer's instructions about feeding. When only a few pots are involved this problem is easily overcome by almost filling the pot with potting compost, making a 1" wide x 2" deep/2.5 x 5cm hole in the compost and filling this with seed compost. The seeds germinate in this area and by the time the roots reach and need nutrients it is there lower down the pot without any further potting up required.

Truss size

Total crop weight depends on many factors, one of which is the size and earli-ness of the bottom trusses. Many varieties respond to their potential to produce large or double 1st and 2nd trusses when grown as follows: the seedlings are grown for the first week after germination at a daytime temperature of 68°-70°F/20°-21°C, then for the next 7 days the temperature is dropped to a consistent 55°F/12.5°C day and night. Sometimes this is no problem since it fits in with the weather pattern. Then it is returned to 65°-70°F/18°-21°C day and 58°-60°F/14°-15°C night during good light conditions, a few degrees cooler in prolonged dull weather. It is during this critical period that the flow-ering potential is being determined in the growing point and, although microscopic, the flower number is affected by temperature. Plants grown at

60°-65°F/15.5°-18°C day and 55°F/13°C night are sturdier and have fewer leaves before the first truss, but they take longer to grow and although the first trusses are heavy, they take longer to ripen. When a cold crop (one grown without any artificial heat once planted) of 3 or 4 trusses is being grown this method will produce a heavier weight of fruit per plant and a heavier total crop weight. For a full crop, tomato plants raised at a constant daytime temperature of 68°-70°F/20°-21°C grow quicker, taller and produce more leaves before the first truss. Depending on variety they usually have a single truss of 6-8 fruit which ripen earlier. Not only do they produce earlier fruit but with a lighter bottom truss, the plant establishes a steady pattern of growth and cropping over a long period.

Summary

•Check germination temperatures to be as accurate as possible.
•Look for signs of growth after 4 days: some F1s are quick to germinate.
•Allow up to 3 weeks for old varieties to come through as some are erratic.
•Do not allow germinating seed to dry out.
•In late spring protect emerging seedlings from strong midday sunshine.
•Carefully remove unwanted direct sown seedlings with minimum disturbance.
•Discard any malformed seedling or any that are reluctant to shed their seed coats.
•Prick out thickly sown seedlings just as soon as they are large enough to handle.
•Pot up space sown seedlings as soon as their leaves start to touch and shade.
•Sow some extra seed to allow for choice of seedlings.
•Accurate temperature manipulation can influence the size of the first truss.

9 Propagation

One of the joys of tomato plant propagation is to see a bench full of sturdy dark green hairy plants; it is a most rewarding and lovely sight. All the potential for a delicious heavy crop is right there on the bench. A well grown plant is the result of paying attention to the many small details that eventually become second nature to the tomato grower yet together they add up to successful tomato propagation.

Until Mr W. J. C. Lawrence and Mr J. Newell of the John Innes Horticultural Institution formulated the standard for 'John Innes Seed and Potting Composts' in the late 1930s, every gardener had their own recipe for seed and potting composts. The nutrients in the John Innes composts are based on hoof and horn, superphosphate of lime, and sulphate of potash. As the last two are artificial fertilisers they are unacceptable or only for restricted use in an organic growing system which more gardeners than ever wish to follow. Good loam is the basis of John Innes compost, but as the supply of loam could not meet the demands of an expanding horticultural industry in the late 1950s-1960s, an alternative had to be found. Peat became the new compost base, a sterile, water holding, easy rooting natural material. Mixtures containing peat plus sand or grit, perlite or vermiculite, together with chemical nutrients, became the preferred compost for both commercial and amateur gardeners alike.

Loam alternatives

When it was realised that peat reserves were being rapidly depleted, yet another alternative base was sought. Coir, a by-product from the husks of coconuts, featured prominently in several composts for a time, but many gardeners found its uneven water requirements difficult to cope with.

Straw, shredded and composted wood and tree bark, spent brewers grains and hops, mushroom compost and all types of good farm manure and composted green waste have been tried in a whole range of mixtures. In the right proportions most of these materials will make good soil conditioners, but few are in a form which meets the requirements of the organic tomato propagator. The widely variable and inconsistent results from most of those early mixes show just how consistent the original John Innes Composts are and just how difficult it is to try and replace them organically.

During the last decade, many attempts have been made formulate a compost from a mixture of a peat-free base and suitable nutrients in both organic and inorganic form and there is now a range of composts which are being constantly improved. Whatever type of compost is used each has its own individual character. Soil-less types tend to hold water longer and often produce a softer plant than a soil based one; some of them appear dry on the surface but are wet enough lower down. If there is any doubt pick the box or pot up and feel the weight of it: a dry compost is appreciably lighter in weight than one that is surface dry.

In the days when all plants were grown in clay pots, the watering can was held in one hand and a cane with a small boxwood mallet head in the other. The pot was tapped and if it responded with a hollow ringing sound the plant needed water. Plastic pots do not respond to such musical chimes and need careful watering to keep uniformly moist.

Vital oxygen

Oxygen in potting composts is all important. Roots grow by cell division, the energy for this comes from respiration. Nutrients are absorbed in the area behind the root tip, absorption also needs a lot of energy, which is also provided by the root respiration, this is powered by sugar transported from the leaves. Water is taken in by the fine root hairs; their development depends on the physical texture of the compost, its aeration, temperature, moisture and nutrient content, and pH. As all of these affect the speed of root growth, they are well worth a closer look.

Texture and aeration

The texture of soil based compost varies with the sand/clay/fibre ratio. A sandy, fibrous, low clay faction soil is quick to respond to temperature changes, it is free draining and contains plenty of oxygen. The high clay, low sand and fibre type on the other hand needs sharp sand up to 3mm grist adding to it to aid drainage and aeration.

During potting it is all too easy when using a heavy potting compost to ram the thumbs down into the pot and create a dense air-excluding mass. This completely closes up the soil pores, those vital air-filled spaces in the compost. If the air-filled porosity of the compost in a pot drops below 10-15%, growth rates slow down and roots start to die in the lower levels of the pot, especially in a heavy clay soil.

Soil pores contain more carbon dioxide than the surrounding air because of the root respiration, this needs to be replaced by oxygen from the air. As water drains through non-capillary channels in the compost, air is sucked in and replenishes the oxygen supply to the roots. Capillary pores cannot be drained by gravity but by suction, and the narrower the pores the greater the suction needed. This is one of the advantages of a solid sand filled bench. When the size of the sand grains is similar to those in the compost, they exert a drainage pull and air follows. If the sand bench is also heated, a gentle radiant heat surrounds the plant and excellent root growth rapidly takes place.

Root temperature

It is as well to bear in mind, especially at the propagating stage, that tomato plants will not grow new roots at temperatures of less than 57°-58°F/14.5 °C. Instead they remain static until the temperature rises. If they are exposed to low temperatures for long periods, not only is growth checked but if planted out into cold soil, the plants are susceptible to attack from any parasitical fungus spores or bacteria that may be lurking around. Small soil thermometers are not expensive and when pushed into pots or boxes on a bench, show the present temperature at root level. A maxi-mini thermometer placed on the bench at leaf height will record temperature extremes where it matters, at plant level.

Plant nutrients, and pH

In winter a tomato plant may be in its pot for 3 months, slowly absorbing the nutrients from the compost. Come spring it may only take 4-7 weeks before the food reserves in the compost are exhausted and liquid feeding with dilute tomato feed is needed before planting out begins. Yet potting compost is the right medium at this age, for if seedlings are potted into a final growing compost too early the concentration of the nutrient solution in the compost will be too strong and the fine root hairs will be unable to develop properly. The overall balance of nutrient release is conditional on the pH level.

The letters pH stand for **p**otential **h**ydrogen ions which are used as a measure of soil acidity. It is a logarithmic scale with pH 7 representing neutral. Below 7 is acid and as it progressively decreases to low pH values aluminium and manganese salts along with other elements become increasingly soluble, poisoning the plant. Above pH7 indicates an alkaline soil solution which 'locks up' several elements by making them insoluble and a deficiency occurs, often leading to a yellowing of the leaves called lime induced chlorosis. Tomato plants will grow across a pH range but are better grown between a pH of 6.0-6.75 when most major and minor trace elements and nutrients are available to the plant. Simple soil-testing kits are readily available containing full instructions and a guide to any lime required.

Watering

The amount of water needed during propagation depends on the size of the plant at each stage and very much on the prevailing weather. It is better to fill the pot to the rim and allow it to drain, absorb air and reach the moist side of dryness before rewatering than keep it constantly wet. Research has shown that the temperature of the water has no significant effect on overall growth.

The research was conducted in a heated greenhouse; pots watered with mains water lost heat but within half an hour they were back to the ambient temperature. On the other hand, some gardeners argue that when they are growing in cold structures, sunlight determines the temperature; during a cold spell the soil barely reaches rooting warmth and growth stops. Rather than run the risk of checking the plants by adding even colder water when there is no hope of immediate sunshine, they prefer to keep a can of water in the greenhouse so at least it is at air temperature.

Pots: clay or plastic

Glorious spring sunshine; as it floods over the propagating bench, tomato plants positively soak up its life promoting light. To take full advantage of every sunbeam, the seedlings need to be behind clean glass in a good moist compost and in the right size and sort of pot. But what is the right pot? Large or small, clay or plastic? The resultant success of the plant depends on the pot used. Both the early and total crop yield (all other things being equal) are determined right there on the propagating bench.

The traditional clay pot, which needs scrubbing both inside and out before use each year, still has its devotees. Water evaporates from all the surface area of the unglazed clay, this cools the compost and it dries out quicker than in a plastic pot but it does allow more air to enter the compost and promote a vigorous root system. The cooling effect reduces the temperature of the compost by some 6°F/3.3°C during the day and 2°F /1.1°C at night compared to compost in a plastic pot. Weight is another factor: plastic pots are appreciably lighter to handle and are much easier to wash.

With the introduction of better structured potting composts, the time honoured practice of crocking the pots became unnecessary. Indeed researchers found that crocks could actually impede drainage, particularly on a sand bench. The layer of crocks in the pot bottom provided an effective barrier to the drainage pull of the sand, which is exerted when it is in close contact with the non-capillary pores in the compost contained in the pot. Whether clay or plastic, the best drainage and aeration is provided when 25% of the base of the pot consists of holes.

In the past 'potting on', that is repotting into a larger pot when the plant has become root bound, was a practice that owed more to producing ornamental plants than tomato propagation. It probably arose from the need to frequently replenish the low level of nutrients then present in potting soil and may have originated from the fact that many ornamentals are better when kept moving up a size. Research has shown that the root system of a tomato plant is invariably damaged during potting on, with a subsequent and cumulative reduction of the early yield every time it happens. So the less the root ball is disturbed the healthier the root system becomes and the better the potential crop.

Types of bench

A solid sand filled bench, heated from below and radiating warmth, produces sturdy plants, but the very opposite can be the result from a cold, waterlogged sand bench in winter, especially one that is built right up to the greenhouse wall. Many modern glasshouse heaters are placed on the centre path. Warm air rises to the ridge and as it cools, falls down the inside of the structure. A solid bench right up to the wall forms a barrier to the descending air stream, which then moves across the bench surface as it is sucked into the warm air rising over the centre path. As it passes over the sand the air evaporates water from the bench cooling it even further and also adding to the humidity.

A soil thermometer in a pot on the outside of such a bench can reveal a compost temperature many degrees less than what is assumed to be the greenhouse level from a thermometer fastened up at head height. The effective growing temperature that really matters is at plant level not what we feel on our faces. This is where slatted or mesh benches produce better plants. A 6"/15cm gap should be left between the staging and the outside wall, so that the cooling air sinks to ground level before it is sucked up into the heated convection current again.

With warm air surrounding the plants on the slats, they do tend to need a little more water as they dry out quicker than on a sand bench, but air follows the water through the pot and the result is a first class plant.

As the plants grow they need more bench growing space, and ideally each plant should stand clear of its neighbours so that the leaves do not overlap and cause shading. On most springtime propagating benches this is easier said than done as fast growing plants compete for bench space. Often a compromise has to be reached between the ideal and the practical. Nevertheless, seedlings and young plants allowed sufficient growing space usually develop into sturdy, dark green hairy plants preferably wider than they are tall with a well developed truss in the head of the stem. When the first flower opens on the first truss of half the plants in the batch, it is time to plant them all in their ultimate cropping borders, containers or pots.

Summary

•Maintain bench temperature above 57°F/14.5°C, which is rooting temperature.
• As far as possible allow each plant sufficient unshaded growing space.
• Time the propagating period to fit in with the date when the greenhouse is likely to be warm enough for pollination to occur.
• Keep the propagating greenhouse glass clean.
• Use only clean trays and pots for plant growing.
• Avoid old or stale potting compost.
• Moisten the potting compost the day before using and bring it into the warm greenhouse.
• Choose a suitable compost.
• Five inch/13 cm plastic pots are large enough to contain a fine sturdy plant.
• Keep pots moist but not soaked.
• Cultivate a 'gentle thumb': it is all too easy to ram compost into a plant pot with

terrific pressure from the thumbs squeezing out much of the air. A firm but gentle pressure gives better results.

10 Growing in the greenhouse border

There are many ways to grow tomatoes, from hydroponics – which are completely soil-less, with the roots being in a constant flow of liquid nutrients – through to rooting into inert substrates like rock wool, which anchor the plant into a recirculating nutrient solution. Then there are all sizes of container, pot and growbag crops, the strawbale and ring culture methods and different combinations of several techniques. Commercial growers are keen to try any method which increases efficiency as the marketplace becomes ever more competitive. But for all the modern innovations home gardeners still look for a simple way to grow their crops. Sound, healthy organic soil culture is still the simplest and easiest way to grow heavy crops of well flavoured good quality tomatoes. Such soils are the very foundations of the crop and need to provide structural conditions for both the tap root and the fibrous root systems.

For generations, tomato plants have been grown directly in the greenhouse border where the roots have access to a large volume of soil. Not only does the soil provide a reservoir of water and both soluble and insoluble nutrients, but it contains essential trace elements and micro-nutrients. It is often been said, but the answer really does 'lie in the soil'.

Tomato roots

The taproot of a tomato plant, if unbroken during potting and transplanting, can grow down some 4–5 feet/1–1.5m into the subsoil. Strong lateral roots may extend up to 3 feet/1m from each side of the stem. A plant that only has this type of root system will be strong and tall, with large widely spaced leaves, bearing long thick trusses, carrying widely spaced tomatoes. Sometimes the fruit is hollow and imperfectly coloured, with a poor flavour, lacking in both eye and palate appeal.

From the lateral roots there grows a fibrous root system of hundreds of short thin feeding roots. A plant with only short laterals and a prolific fibrous root system usually produces a good crop of high quality fruit on 5-6 trusses before the growing point becomes weak and spindly. This is fine for a plant that is intended to be stopped at the 6th truss.

For a full long season's crop of vigorous plants what is needed is a border soil that encourages a framework of long healthy roots to physically support the stem and weight of the plant, together with plenty of fibrous feeding roots.

Root distribution

Most tomato cultivars will grow 60–80% of their root system in the top 12"/30cm of soil, 20-25% in the next 8"/20cm below that, with the rest going deeper. The depth to which a tap root will grow depends on the structure of the soil, the subsoil and the temperature and moisture levels.

Glasshouse tomato plants will not root at temperatures below 57°F/14.5°C. By the time the lower depths of soil in cold greenhouses are warming up, the plant will be already established in the warm upper soil with a good supply of fibrous roots.

Tomato plants are positively geotropic and will grow downwards in search of water, but they are also hydrotropic, which means that they will respond to the presence of water at whatever level. If the water regularly supplied is only sufficient to moisten the top 2-3"/5-8cm of soil, then that is the area where most of the fibrous roots will be found.

One of the joys of growing tomatoes in the open greenhouse border is that they can be all things to all people, from just a few trusses to a space-filling full crop.

The simple way

When only 3 or 4 trusses are being grown, a healthy soil is refreshed with a large bucketful of well made garden compost or old well rotted manure dug into each square yard or metre of bed. A refreshed healthy soil should provide enough sustenance for the season for plants that are stopped after the third or fourth truss has formed. This is organic tomato growing at its simplest and easiest. Plants being grown on into the 6-10 truss range require added nutrition as the season continues, to maintain healthy growth and swell all the fruit on the trusses. These are fed similarly to the long season crop. For a full long season crop when the aim is to grow 12-15 plus trusses of good quality fruit entirely by organic methods, then tomato culture becomes a fascinating challenge.

A full crop

Assuming that the border soil is disease free, then it needs to be
• Firm enough to anchor the plant.
• Water retentive.
• Free draining.
• Have a pH in the range of pH 6.0 – pH 6.75
• Have a good porous structure to permit plenty of air to reach the roots.
• Contain a supply of organic nutrients.
• Contain a range of minerals and trace elements
• Be warm enough at planting time: 57°+F/14.5°+C.

Each soil type has its own characteristics and needs cultivating accordingly, always bearing in mind that under glasshouse conditions the whole life processes of flora and fauna and nutrient availability are speeded up.

Clay soils

Heavy clay soils are usually inherently fertile, but when saturated or baked become unworkable. Such soils may need sharp sand or grit digging or forking into them to improve their physical texture and drainage. Just how much is needed depends on the density of the clay. A test with a pH kit will reveal the pH level of the soil. If the need for lime is indicated, by a pH reading below 6.0, magnesium limestone, also known as Dolomite Limestone, used at the amount indicated will add magnesium to the soil whilst liming it. Where larger amounts of lime are needed(i.e. a reading below pH5.5), a coarse grade of ground limestone will both improve the texture of the soil whilst slowly releasing its alkaline qualities over several years.

The combination of drainage created by the grit and the flocculating effect of the calcium carbonate help to make the soil workable by improving its texture. But the real structure depends on the organic material present in the soil. It is the humic forming potential of compost and manure that combine to release the latent fertility and turn a heavy clay soil into a highly productive substance.

Digging

Initially this calls for hard work as the glasshouse border is double dug. Outdoors lime is applied to the surface of the soil, then as the rain washes it through the soil the neutralising effect takes place. When it is used inside to neutralise and improve the texture, it is easier to dig in the grit and apply the lime first, then a trench is taken out and the soil transferred to the other end of the bed. The bottom spit of the trench is then ' cracked up' with a fork by driving the tines down to their full depth, lifting and cracking up the clod into smaller pieces. It is easy to ignore the condition of the subsoil, but for a long season of plants a good root system is essential.

When fast growing tomato roots reach a stagnant, wet, cold impenetrable subsoil, they invariably die and unless a fibrous root system develops quickly above them, the plant is checked with subsequent cropping much reduced.

Panning

A hard, dry, panned layer on the surface of the subsoil can be just as crop reducing as a waterlogged one. For when the roots hit the hard pan and find no moisture, they start to desiccate and the growth of the plant becomes unbalanced, coarse leaves are produced and small crops of poor quality fruit follow. Once the subsoil is cracked, a small quantity of coarse compost or manure is worked in, not for nutrition but to physically improve the texture of the subsoil whilst opening it up.

The next spit of soil is turned over onto the subsoil and well made compost and or well rotted manure worked into this spit at the rate of a large bucketful every square yard or a little more if working in metres. When a soil is known to be short of nutrients, this is the time to work in any organic fertiliser that may be needed in the form of a balanced base at the manufacturer's recommended rate.

Regular use of compost and manure will build up the levels of minerals and trace elements, but a heavy crop of tomatoes will use up some of this. If there is any doubt, then work in seaweed meal with its high mineral and trace element content at the rate of a handful every square yard.

Silt soils

Silt soils tend to compact and pan quickly and water may not easily penetrate the subsoil, leaving the top spit difficult to drain. They not only have a weak, unstable structure, but any limestone applied will correct the pH whilst failing to have the flocculating effect it has on clay. This makes silt soils one of the most difficult of all soils to cultivate. The coarser silts respond to organic matter being dug in on a regular basis, allowing for up to 50% more than on a clay soil.

On the finer silts, in addition to compost and manure, straw walls may help to increase drainage and admit air. A trench is opened one spit deep at one end of the border. The subsoil is cracked up with a fork by driving the fork in and moving it back and forth, and then a spade is used with the same movement to create a V shaped slit. Straw is laid across the slit and the spade is used to force it down into the slit which leaves the straw sticking vertically upright. This is repeated at intervals of about a foot/ 30cm along the trench. The next spit is then turned over into the trench and compost /manure worked into this. On a very badly drained silt, additional straw walls are made halfway between plants in the top split. In the course of the season there will be some nitrogen loss as the straw decomposes and should this affect the plants it is allowed for in later feeding and top dressing.

Loams

Heavy, medium and light loams are a pleasure to work, being the most fertile of soils. Heavy loams border on the clay, whilst light loams are often referred to as being on the sandy side. Refreshed with a large bucketful of compost or manure forked in late winter, they are the easiest soils to work.

Sandy soil

Sandy soil is easy to work but often not easy to manage. Water drains rapidly through sandy soil, often taking soluble nutrients with it. Calcium, when applied to soils, works downwards, so sandy soils need an annual or biannual pH test which often results in a regular application of Dolomite Limestone. Some sandy soils also tend to form pans but in this case double digging is not just to break up the pan, but also to dig in as much moisture retaining material into that bottom spit as is available.

It is a waste of good compost to put it in at this level but if supplies of low nutrient cloddish straw manure are available this is a good place to put it. On my light sand I have regularly dug two-year-old leaf mould into the subsoil, which is now not dissimilar to the top spit in texture. Once again compost and/or manure is worked in to the top spit of the bed.

High organic content soil

The aim of organic soil management is to feed the soil, which in turn feeds the plant. A combination of garden compost, manure, green manures, nitrogen fixing plants, crop root remains, rotations and fallows, create highly productive outdoor plots. On such soils strong healthy plants, with a high degree of natural pest and disease resistance, regularly grow bumper crops of delicious produce. As the nutrient release is bacteria and fungi powered, it starts slowly in the spring, increasing when the soil warms up as the temperature increases and day length grows, to provide nitrogen, phosphate, potash, minerals and trace elements in keeping with the needs of the developing plant.

Under glass the whole process speeds up, the greenhouse border temperature rarely falls to outdoor levels. Springtime warmth is up to 3 months earlier in heated greenhouses, whilst even in cold ones the soil bacteria is active long before there is any movement outdoors. The time lag between digging in bulky organic material, and the nutrients it contains beginning to be released to the plant depends on temperature and soil moisture.

Garden compost

Garden compost is as variable as its contents. Well made compost should be rich in nutrients and minerals, but the end product is only as good as the materials that went into it. Humus – that often misused word – is the result of the decomposition of ligno-cellulose and animal matter. Unless a compost heap has had a proportion of straw, strawy stems or other ligno-cellulose in its makeup, whatever else it contains it will be low in humus and thus nutritious humic acids. Humus when seen extracted in its final form has the dark brown colour and texture of a jar of yeast extract, it is in a colloidal state with the particles darkening, sticking and binding the soil together. Over time the humus slowly releases its humic acids which are a natural plant food.

When a good sample of farmyard manure is incorporated into a heap or used as an activator, it contains on average 0.6% Nitrogen: 0.3% Phosphate: 0.7% Potash.

When dried poultry manure is used, the nutrient content may on average be around 4.2%N: 4.3% P:1.6%K, which is an appreciable increase and if available is a good activator, but it lacks the straw content of manure and thus the potential humic acids. Deep rooted plants draw up a range of minerals from the subsoil and different weed species contain different amounts of trace elements, which all add to the mixture.

A covering of comfrey leaves added to a compost heap both activates it and increases the nutrient level. From this it is easy to see that a mixture of nutrient-rich materials will result in a compost well able to sustain several trusses of tomatoes. Conversely, a heap of poor materials will result in a compost which may well make a good soil conditioner, while lacking the nutrients to feed more than the odd truss of fruit.

Farmyard manure

The term 'old well rotted manure' is also open to wide interpretation. In many districts, good farm manure is hard to find but may well worth seeking out if a good mixture that has contained plenty of straw can be found. Often horse manure is available from local riding schools, either by the bag or load. Generations of gardeners, myself included, have forked and shovelled tons of horse manure, which was the only manure bought in, all others used to be free for carting. Where the animals are bedded on wheat straw or bracken this is still a wonderful by-product. If it is available, it should be stacked in a rectangle and trampled firm as it is stacked into a heap with sloping sides, covered with a sheet of tarpaulin or polythene held down with stones or timber and left to age for several months until the straw readily breaks down into short pieces when forked. All too often nowadays the horses are bedded down on wood shavings. If the shavings are thin and few and the droppings plentiful, then stacked as before it will eventually break down into useful material. If the shavings are thick and plentiful, leave it well alone as there will not be enough nitrogen present to break the cellulose down and severe nitrogen robbery from the soil will occur. So, as with compost, a good manure is a very valuable asset whilst a poor one may prove to be a liability.

Greenhouse soil temperature

Tomato borders are prepared for planting some 2–3 weeks before planting time. This allows the soil to settle down and warm up, as roots in contact with cold soil will not grow and the plant is checked. At this stage they are vulnerable to attack from any disease spores that are lurking around.

Ten days to a week before planting, plunge a soil thermometer 6"/15 cm into the soil or carefully bury an air thermometer to the same depth. If the temperature is above 57°F/14.5°C all is well. If not, draw the soil up into ridges to expose a greater surface area to the warm air, then rake it down again and gently firm it a couple of days before planting. At planting time the beds should be neither wet nor dry but moist. A warm, moist, thoroughly prepared border is the very foundation of a long season's crop of good quality tomatoes, whatever the greenhouse soil type may be.

Summary

• Recognise the soil type and prepare it accordingly.
• Check the pH.
• Crack up any soil pan.
• If needed improve the drainage.
• If needed improve the subsoil.
• Incorporate the best compost and/or manure available.
• Add organic base fertiliser if needed for a long crop or the soil fertility is poor.
• Create a warm, moist soil condition at planting time.
• Check the soil temperature.

11 Plant spacing

The spacing of tomato plants in the glasshouse border is often a compromise between available space and the vigour of the variety. Research at the Horticultural Research Institute has shown that well managed plants trained as single cordons at the right spacing produce the heaviest yield of good quality fruit.

Broadly speaking and allowing for varietal variation, the closer the plants, the smaller the fruit, but because of the total number of plants, the total crop weight is similar to that obtained from larger fruit on fewer plants. To pack a small greenhouse with vigorous cultivars will create a season of problems, yet to widely space compact varieties in a large glasshouse is a waste of a valuable facility.

Compact varieties with small leaves can be planted 14"/35cm apart in the row, with rows 15"/37cm apart. Most medium sized cultivars are happy at 18"/45cm up to 24"/60 cm between plants, whilst vigorous types need 30 "to 36"/75cm-90cm of space to give of their best.

What matters most is the total volume of the plants when fully grown. Will each plant have enough light and air space? Will warm air be able to circulate freely? And is the structure strong enough for the total crop weight? When the first tomatoes start to ripen there will be several trusses of developing fruit plus the weight of stem and leaves on each plant. This can amount to 10-16lb/5-7kg plus per plant, or approximately 1 cwt/50kg of downward pull for every 7 to 11 plants. To see a superb crop of tomatoes broken or bent on the greenhouse floor for want of better support is a really sickening sight. The most economical and easiest way to grow tomato plants is to twist them up strings fastened to the base of the plant with a large figure of 8 knot and tied to wires suspended over the row of plants. Where the greenhouse roof is strengthened with cross ties, the wires are supported by the cross ties.

In lightweight aluminium houses, supporting the crop weight from wires fastened directly to the roof is to invite trouble. Stout supports at the end of the rows will act as anchors with wires stretched between them to which to tie the strings.

Bamboo canes

In many greenhouses bamboo canes are the preferred plant supports. Medium to heavyweight canes long enough to allow for at least a foot /30cm in the ground are pushed into position before planting starts. As the plants grow they are tied to the cane with soft twine or the short plastic enclosed wire stem ties. Plants carrying several heavy trusses of fruit and secured only by loops of twine or stem ties have a habit of sliding down the cane until the weight buckles the stem and often breaks it completely. When this happens a half broken stem or one that has been badly kinked can be repaired and the plant saved. Lift the plant back up to its former position and secure, replace the edges of the break together as near as possible and bind with plastic insulating tape from below to above the damaged cells. An otherwise healthy tomato plant will droop for a few days, then as the two cambium layers start to unite the plant will start to recover; it does help to remove several of the bottom leaves to reduce transpirational loss during this process.

To avoid plants slipping, tie the soft twine with a knot above a node on the cane before passing it round the tomato stem. When a stem tie is used tightly twist it once round the cane above a node, then take round the stem before twisting the ends together.

The easiest way of all of growing tomato plants is up the metal spirals that have become available in recent years, these are available in different lengths with the 6'/2 metre size ones ideal for tomatoes. Spirals are available in galvanised heavy duty wire form and also coated in green or black plastic; they vary in price and at around £1 to £2 each mean a considerable initial capital outlay but they should last for many years of trouble free growing.

The spirals are pushed into the bed before planting, and the top of the plant is guided into the centre of the spiral once a week and that is all there is to it. Leaves grow out supporting the plant and trusses form, hang and ripen without the need for tying, all that's needed is for the side shoots to be removed when they are small.

Summary
- Choose the right spacing for the chosen variety.
- Fit wires for string supports before planting.
- Place canes in position before planting
- Tie twine above a cane node to prevent slipping
- Consider the initial cost of spirals, balance this against the cost of canes and string, ease of use, and time saved.

12 Planting time

When tomato plants have a truss with the first flower in bloom, or at least showing colour(flowers opening) on 50% of the plants on the bench, it is time to plant them all. Having checked that the soil temperature is above 57°F/14.5°C, during the day, the pots are placed in the final planting positions 2-3 days prior to planting. This gives the plants time to acclimatise themselves to their new temperature and light levels and they will spread their leaves and appear to grow overnight.

During the day or two before planting, give each plant a pot rim full of liquid seaweed diluted to the manufacturer's recommendations. This will ensure that each plant is charged with a full complement of minerals and trace elements, which act as a bio-stimulant to get it off to a flying start. Using a trowel, a hole the size of the pot is taken out of the border soil and, with the fingers across the pot rim to support the plant, the pot is inverted and with a slight tap the roots should slide out of the pot.

Take a close look at the roots: are they healthy, white and plentiful? (The signs of a healthy plant and sound propagation). Should the roots look brown or wizened or should there be any sign of a black, brown or reddish lesion at the point where the stem meets the soil, do not plant it.

A 'clean' potting compost produces 'clean' plants, but if a home-made compost has been contaminated with *Phytophthora parasitica*, a dark brown or black lesion, with a moist diffused leading edge, will encircle the stem at soil level and kill the plant. Sometimes this may appear in its wet brown form. When a plant is dying at soil level with a clearly outlined, dry brown discolouration, the culprit is likely to be *Rhizoctonia solani*. If any form of these collar rots is seen, dispose of the plant, preferably by burning.

Place the plant in the hole at or about soil level and draw the border soil round the root ball and gently firm into position. The root ball should remain at or very slightly above soil level and not covered with border soil, which may contain pathogens.

A mature tomato stem is a very strong tough form of plant tissue, but at planting time its soft stem is unable to withstand a bad attack of soil borne fungi, should it be carried on the water film of the soil and washed onto the stem.

Once planted, give each plant 1 pint/600ml of water carefully round the root ball and for February and March plantings resist all temptation to water again for one week. Instead, on bright days top spray to run off point at midday.

During April, May and June, after the initial watering, more may be needed after 3-5 days. The object of restrained initial watering is to encourage the plant to root out into the border. A week after planting, select a typical plant and carefully scrape away some soil at the point where the root ball joins the border. There should be new white roots growing out into the greenhouse border soil from all around the root ball, indicating that the plant has settled in and is growing. Under ideal conditions these new fine roots may be up to 1½"/4cm long

The young tomato plant

A greenhouse full of young tomato plants growing steadily is a most rewarding sight in early spring. Rewarding for the satisfaction of seeing the plants from seed through propagation to being established in their cropping position. Rewarding financially too. When a tomato plant receives a check to its growth, either by temperature extremes or water stress, prolonged shortage of light or nutrients, there is a reduction in the final crop weight. These reductions, although unquantifiable, are accumulative. The more a crop is checked, the less the total yield.

Throughout the life of the plant, from seedling to season's end, the aim is to keep the plant growing steadily.

Growing temperatures

Steady growth is maintained when the speed of root expansion keeps pace with the leaf and growing point development. Once established, the roots should not be allowed to dry out, or become waterlogged, but kept moist with the quantity of water applied in keeping with the stage of development of the plant.

Leaf and plant development depends on photosynthesis, which in turn depends on the quality, intensity and duration of light falling on the leaves, preferably within the temperature range 60°-70°F/16°-21°C. So much research time has been devoted to tomato growing temperature and the result for commercial growers is a tomato growing blueprint. Where fully automatic heating and ventilation systems are used, precise levels of 68°F/20°C during the day and 58°F/15.5°C night are maintained during good light conditions, with temperatures allowed to rise during summer ripening conditions.

Without the benefit of automatic equipment, precise temperatures become academic, but what can be done is to grow the plants as far as possible within their natural temperature range with the facilities that are available.

Soil temperatures above 57°F/14.5°C and air temperatures as above should be the aim during good light conditions. Balancing warmth and ventilation during the fast moving weather conditions of a late spring day can be tricky. When the glasshouse temperature is too hot for too long, pollination and fertilisation may be badly affected. This is when automatic ventilators are a real boon in a small greenhouse. Set to rise at 72°-74°F/22°-23°C, they are a cheap way of temperature control. Manually lifted vents offer immediate cooling, but need regular attention during rapidly changing conditions.

Growing in pots

When the greenhouse soil is unfit for tomatoes, or there are no soil borders or it is too cold for early growing, good crops of excellent quality fruit can be grown in pots. The big advantage of pot grown plants is their manoeuvrability. Early in the year they can be stood close together in the lightest, warmest part of the greenhouse gaining days, or even weeks of growing time before being moved to their final quarters.

Dwarf plants will grow in a 9"-10"/25cm pot. Early outdoor varieties such as Red Alert are well worth growing inside in pots as they will provide some early fruit long before the main crops are mature. These dwarf plants can also be grown in baskets and hung wherever there is space to hang them; under glass they tend to ripen over some 2-3 weeks, then the crop is over, but what a welcome to the new season.

Cordons grown to 8 plus trusses need a minimum of a 12"-15"/30-38cm pot. Recycled 3 gallon/13.5 ltr buckets or containers, with holes drilled into the base, allow a good root run and space for top dressing

Growing compost

Home made potting compost, or a rich organic soil give good results when used with top dressings and liquid feeds. For anyone unable to make their own, there are now a good range of potting composts available, which produce good results. Pot grown plants need watering at least once a day but otherwise their culture is the same as for border grown plants. Because of the limited volume of compost in the pot, an occasional application of liquid seaweed at the manufacturer's rate of dilution, both as a root bio-stimulant and a foliar feed will top up the supply of minerals and trace elements needed by the plant. The plants are invariably supported by canes, and although the pots may appear cumbersome and heavy when wet, they can be placed in the best growing positions and moved as necessary.

Grow bags

The rise in the popularity of grow bags reflects their versatility and ease of use. With careful watering and feeding they can produce remarkable crops of a wide range of tomato varieties. The shallow depth of the bag allows it to be used for both tall crops or a short 'catch crop' where the headroom is so limited that even a pot would restrict the height of a plant to be grown. Bag size and the small volume of compost it contained used to be a restricting factor but now larger bags are available. When growing in a large capacity organic grow bag, the volume of compost allows a larger root system to develop whilst the available nutrients get the plant off to a good start.

Watering and feeding are the most important parts of bag culture. If the bag is allowed to dry out blossom end rot almost always follows. Overfeeding the plant with too strong a feed with the resulting strong nutrient soil solution may also induce blossom end rot when the water level in the bag is low without it actually drying out. Because the plants look small there is often a temptation is to overplant grow bags. Much better results are to be had from 2 or 3 plants per bag than with 4. Warm air movement can be restricted by leaves hanging down over the bag so leafing (see p.45) should be done as with border plants.

Summary

- Check soil temperature is above 57°F/14.5°C.
- Day before planting, water plants with liquid seaweed.
- Plant when 50% of plants have a flower open.
- Water in with 1 pint/600m per plant.
- Top spray daily for the first week unless the weather is wet and the air damp.
- Check the health of the plants, discard any showing signs of disease.
- Aim to keep day temperature in the 60°–70°F/16°–21°C range.
- Aim for a night temperature of around 58°F/15°C

13 Watering

Of all tomato growing operations, to water or not to water is the question most gardeners frequently ask themselves; so simple a question yet so important is the answer that, assuming the rest of the plant management is satisfactory, the quality and quantity of the entire crop depends on getting it right.

Why? Because the very life of the plant depends on water. As the transpirational stream flows up the xylem tissue, it distributes water and nutrients throughout the plant. This is powered by the transpirational pull, that continuous loss of water vapour through the stomata in the leaves. Replacing this with just the right amount at the right time is one of the most important skills of tomato growing. Knowing how much to give and when only comes with experience of growing plants under your particular conditions. Fortunately the guide lines are the same for all.

Cooling the plant

On a hot summer's day a tomato plant is cooled by the evaporation of water from its surface. Around 5% is lost from the epidermal cells, but over 90% is through the open stomatal pores on the leaves. Water loss is most rapid when a fully turgid plant has absorbed heat from the sun, is in a very dry atmosphere, with hot dry air moving quickly over the plant. Under these conditions even a healthy plant cannot transport water quickly enough up the stem to meet the transpirational demand, and wilting can occur. Usually the plants recover when the heat goes, and/or the humidity increases, either later in the day or during the night.

In severe cases, the water columns in the xylem are disrupted by the loss of pressure and cavitation takes place. This is the formation of vapour filled cavities in the mainstream of the stem. Plants dying of drought in hot summers either inside closed greenhouses, or in outdoor parched soil are often found with shrivelled dry stems where the cells had collapsed before the water column could be refilled.

Tomato plants grown permanently short of water are invariably small for their type, not only in height, but their leaves are smaller and the truss size restricted. The number of fruits produced is less and because of their size the yield is low. But the flavour! The flavour is usually superb, this is due to the low water content of the soil increasing the soil salinity. This can be used to good effect when growing cherry tomatoes. Researchers at H R I found that Gardeners Delight grown in soil and kept on the dry side of moist had a much better flavour than when grown wet.

Dry growing

The balance between 'dry' growing to create a good flavour and poor setting and blossom end rot is a delicate one: sufficient water is needed to maintain the plant; when plants have been kept short of water at the fruit swelling and ripening stage, they are small with thick skins.

Under these conditions care is needed not to suddenly saturate the soil with a heavy watering, as this is invariably followed by the ripe fruit cracking or splitting as the thick skin is put under pressure from the excessive amount of water suddenly available.

Dry growing is the exception, all too often tomato plants are overwatered. The very weight of full watering cans tends to regulate their use, but it is all too easy to be generous with a hosepipe and give the plants a 'good' watering. A large lush plant bearing strapping great leaves with trusses carrying tasteless fruit is a typical example of enthusiastic overwatering.

On average, most summers have wet humid spells during which transpiration is either very slow or stops altogether. When soil, plants and air are saturated, don't make things worse by watering. Keep the tap shut until the weather improves.

Quantity of water

Once planted in the bed, tomato plant needs about 20-22 gallons/91-100 litres of water during a full season. As much again is lost by drainage and soil surface evaporation, so around 40-44 gallons/182-200 litres per plant per full season is a guide. Short season plants need about half this amount

When the following table was being developed at the then Fairfield Experimental Station in Lancashire some 45 years ago it suited their free draining soil. At the time, those of us growing on organic soils found it too much. Today it is still a very useful guide for most soils, but when tomatoes are grown in a high humus organic soil, which is then mulched, they need far less water than when grown in a free draining mineral type of soil.

Weather	Water needed by plant for a 24-hour day	
	Pints	**Litres**
Very dull, dark and cloudy for most of the day.	¼ to ½	0.14 to 0.28
Dull, cloud overcast most of the day.	½ to ¾	0.28 to 0.42
Fairly sunny, cloudy with bright periods	1¼ to 1½	0.71 to 0.85
Sunny only an odd cloud	2 to 2¼	1.10 to 1.20
Very sunny, clear sky all day.	3 to 3¼	1.50 to 1.80

The water-holding capacity of soils varies so much that it is worthwhile checking the water status every 14-21 days. Choose a spot of the border that is halfway between plants and take out a small trowelful of soil to a depth of 8"-10"/20-25cm and squeeze it in the hand. The border soil should be moist enough to bind together with just a little water being squeezed out, if it needs more or less adjust accordingly.

Summary
- Consider today's weather and the present weather pattern.
- Allow for the age of the plant.
- Water according to the type of soil.
- Consider the water holding capacity of the border.
- Check the moisture level of the soil.

14 Pollen and pollination

Every complete tomato fruit begins life with a speck of pollen being transferred from the anther to the receptive stigma of the flower. The natural form of most glasshouse cultivars with their 5 or more anthers hanging down close to the stigma is ideally adapted for self-pollination and indeed from mid-season onwards this is what usually happens. Whether a fruit forms or not depends on these 6 words: viable pollen, receptive stigma and pollen transfer.

Pollination temperature

Although pollination of glasshouse tomato varieties can take place over a narrow range of temperatures, the optimum for most of these is 68°F/20°C for both anther and stigma alike.

Research evidence suggests that the actual potential quantity of pollen produced by a flower depends on the variety and the intensity of the light levels when the truss is first initiated in the head of the plant. The actual quality and potential viability of the pollen will depend on the nutrient status of the plant and health of the flower itself.

Tomato flowers

Tomato flowers close up at night and the speed of opening next morning depends on what type of day it is. Usually by mid morning on a bright day they are wide open and if it is warm enough they are producing pollen.

Once a tomato flower is fully open and mature, it is at its most fecund over the next 3 days. Very early in the season this gives a 1 in 3 chance of successful fertilisation.

At the other end of the scale, bright sunshine for long periods in an unventilated greenhouse can easily raise the temperature to over the 100°F/37.5°C mark. Four hours at a temperature of 105°F/40°C will cook and kill the pollen.

Wet and dry atmosphere

When the greenhouse atmosphere is too dry the pollen will not stick to the stigma, but if the atmosphere is too wet the pollen clings to the anthers and will not 'fly'. A relative humidity of 70% suits both the anther and stigma and ensures a good set.

Early in the season setting those first flowers is all important not only for the pleasure of enjoying those early tomatoes but also to 'anchor' the plant when it is growing vigorously in the improving daylight. Each day has to be judged by the weather conditions: on a bright but cool spring day it may be necessary to close the ventilators for half to one hour during late morning. Around noon top spray the plants either with a coarse pressure spray, fine rose on a watering can, or a finger over the end of the hosepipe. Spray just long enough for the water to start to run off the ends of the leaves. This will increase the humidity. Return an hour later, open the ventilators in keeping with that day's weather and tap the wires or canes with a stick, or when only growing a few plants, gently tap the trusses in flower. On a good day the pollen can quite clearly be seen to fly, fertilisation takes place and those first fruits start to form.

Top spraying the plants

Hand spraying the plants with water to improve the humidity is now rarely practised in commercial tomato growing. For large production units, exact control of the glasshouse temperature and humidity is carried out automatically. Fertilisation in large blocks of glasshouses is done by bumblebees. Small colonies of bumblebees, *Bombus terrestris,* in cardboard hives are hung from the structure and the bees fly around and fertilise the flowers as the pollen is formed. When wandering bees find their way into small greenhouses don't dissuade them as they too will be busy spreading any pollen that is about.

Top spraying should not be done when it is too cold, when it is raining or when the atmosphere is already humid. If it has to be done later in the day, do allow time for the plants to dry off before nightfall.

As the season warms up the vents are no longer closed for an hour and by the time the plants are 3-4 trusses tall self-pollination becomes more effective. On buoyant days it is still worthwhile top spraying and/or tapping, as this can make all the difference between a medium and a heavy crop, or between profit and loss.

To check if the pollen has been viable and fertilisation has taken place, gently pull a spent flower from the calyx; is there a tiny round fruit to be seen? are most of the spent flowers like this? If so, that batch of flowers has set and the swelling process has begun.

Dry set and parthenocarpic fruit

If, due to low light, high temperatures or water stress, no pollen is produced or is non-viable, dry set occurs and the flowers shrivel and dry up on the truss. Occasionally when a plant is carrying several very heavy trusses of fruit it misses or 'skips' setting a truss, and only resumes making pollen when some of the fruit has been picked and the plant is recovering its strength. Parthenocarpic fruit are the little tomatoes found at the end of long, or partially set trusses, they are a feature of some varieties and as they were not fertilised they contain no seeds but if they grow large enough are often sweet to eat.

Summary

• Bear in mind. No pollen = No fertilisation = No tomatoes.
• For optimum pollination aim for a temperature at or near 68°F/20°C.
• Aim for a relative humidity at of near 70%.
• Balance temperature and ventilation early in the season.
• Tap wires or canes, or gently tap the trusses early in the season to distribute the pollen.

15 Training and leafing

By nature the tomato plant is a sprawling vine; it will grow, fall over and continue to grow horizontally along the ground or up and over anything at ground level or at a low height for as long as the growing conditions continue. Bush varieties planted in the open ground without any further attention can grow into a large mound of stems, leaves and trusses of fruit. Cordon varieties supported within a frame will turn into a mass of luxuriant growth as every side shoot becomes a stem with its own leaves and trusses. It is possible to leave plants alone and get a crop, but more often than not they turn into jungle of vegetative growth at the expense of fruit.

Extensive research has shown that tomato plants trained as single cordons, planted at the right spacing for the variety, will yield the heaviest crop of good quality fruit. In practice a well trained tomato plant is:
   • Easier to keep healthy
   • Easier to manage
   • Quicker to train and trim
   • Easier for the fruit to ripen
   • Easier to pick
   • Easier to observe the health of the plant, or to 'read the plant'

Side shoots

Side shoots grow out of every leaf axil, just where the leaf grows out from the stem and should be pinched out just as soon as they can be handled. Once a shoot is over 2"/5cm long, it starts to reduce the potential total yield and, as there is a shoot in every axil, nip them out. The exception to this is where there is not really enough room for another plant in the corner of the greenhouse, but space is being wasted, here a shoot is allowed to grow and then trained up to fill the space.

When growing up strings, once a week the string is twisted gently round the plant in a clockwise direction, taking care not to trap either leaves or trusses, twisting where possible underneath a truss to support it.

Often plants are stopped at the wire, but early crops with a lot of growing time left have now to be retrained. An easy way is, when tying them up at the start of the season, to leave an extra yard/metre of string as each plant is tied to the top wire. When the first two trusses have been picked and the leaves removed up to the next ripening truss, the plant is untied and carefully lowered and retied further along the wire. The stem will sag into the new position but not so low that the fruit on the 3rd or 4th trusses touch the soil. There are several commercial variations of this but this is the simplest, it does help to have someone to assist by taking the weight of the plant during the process until the knack has been gained. Another way, once the tops have reached the wire, is to train them along it. In this case it is imperative that the shoots are removed or a veritable jungle will develop, with any passing pest or disease only too ready to take up residence.

The same is true of the modified 'Guernsey arch' method. Plants grown up string or canes are trained over horizontal wires into an arch over the path between the rows, thus gaining another 3–5 trusses per plant. The growing point is carefully trained over and tied down to the next wire until the arch is formed. Whatever method is used, tomato plants need to have their heads in the light with the growing point unshaded, and warm air able to circulate freely around them.

Leafing

Leafing or, to be correct, the de-leafing of tomato plants is a subject guaranteed to provoke a lively discussion whenever tomato growing gardeners and botanists meet. Indeed the deliberate defoliation of a healthy plant does seem to be a drastic action for any plant lover to take. So taking a closer look at the facts of this contentious subject I will try and put it into perspective.

If the roots are the powerhouse of the tomato plant, then the leaves are certainly the factory. Each season they are so readily taken for granted but they are truly one of nature's miracles. Given warmth and light strong enough to start photosynthesis, they absorb carbon dioxide from the air and turn it into life promoting compounds of great complexity.

New leaves

Under ideal summer growing conditions, a tomato plant will initiate a new leaf every 2 days. In order for it to develop, it first draws nutrient from more mature leaves further down the plant. Then, as it expands, it starts to photosynthesise and for a time may actually import and export nutrient before settling down to become a prime producer of 'assimilate'. Assimilates are the first simple organic compounds that become such a complex mixture of nutrients required by the plant to maintain its life processes. Leaves also store several vital trace elements and macronutrients such as magnesium, which is only noticed when a deficiency is confirmed by the inter-veinal yellowing of the older leaves as the magnesium moves up to the young leaves.

During hot weather, a tomato leaf evaporates water both from its surface and from open stomatal pores to keep the plant cool. In short then sufficient healthy leaves are vital to the well being of a tomato plant. So why cut them off?

The answer lies in the word sufficient. Over the many years of growing wild tomato plants and leaving them to sprawl naturally, I have noticed that the oldest leaves wither and die when they reach a certain age and the plant becomes too crowded and the competition becomes too intense.

Modern greenhouse cultivars have much larger leaves spaced further apart on the stem, but most of the natural traits are still there. Some years ago research scientists at the Horticultural Research Institute found that tomato leaves, given good levels of CO_2, photosynthesised most efficiently when they were about 30-50% expanded or about 1,000 mm square in size. For the next 10 days they were at their maximum efficiency, which then started to decline. By the time the leaves were 32 days old, the rate of photosynthesis was down to 10%.

Leaf curling

During the longest days of the year with widely fluctuating temperatures, there appear thick, old, tough tomato leaves with their florets curling upwards under the pressure of stored starch which has become so solidified as to be virtually immobilised. These old leaves contribute nothing to the plant and in old age start to turn yellow, hang down and die, and would have been far better removed. Even when turgid, old curled leaves provide a safe haven for pests and restrict air flow. Warmth is the key to ripening.

The gentle movement of warm air speeds the ripening process among mature green fruit, and this air movement is hampered by a full complement of leaves.

The amount of leafing practised depends on greenhouse design, variety and spacing. A single row of plants, with an open habit, 30"-36"/75–90cm apart, offer little resistance to air movement, while a double row of large leafed close jointed types at 15"/38cm centres may stop it completely. Leafing little and often does not check the plant and maintains a balance between:

• Photosynthesis and the plant's current needs.
• Stored starch and its translocation.
• Air movement and sunlight penetration.
• Ease of crop management and reduced risk of pests and diseases.

There comes a morning when on opening the greenhouse door the air is stale. That is the day to remove the bottom 2 or 3 leaves. Some 10 to 14 days later, depending on the time of year and the speed of growth, if the same staleness is noted, it is time to remove another couple of leaves. This process is repeated when necessary but only remove leaves up to a ripening truss. In this way there will always be sufficient leaves on the plant to supply its needs.

Leaves are better removed as early in the day as practicable to allow the wound to form a callous before night fall. A clean cut with a sharp knife through the 'collar' of the petiole is all that is needed. When debating whether or not to leaf, bear in mind that, important as tomato leaves undoubtedly are, the very best ones are only a means to an end, what is really being grown is the fruit.

Summary

• It is better to remove leaves little and often.
• Only leaf up to a ripening truss.
• Leaf early in the day to allow for fast healing of the cut surface.
• Use a sharp knife and cut cleanly through the 'collar'.

16 Feeding the plants

Tomato plants are sometimes called gross feeders but the amount of nutrients that they require depends on the type of crop being grown. The organic way of feeding the plant, that of feeding the soil which in turn feeds the plants, works well enough with glasshouse tomatoes when only a few trusses are being grown. Sustaining a tomato plant during a long season of full cropping however is a very different matter. Once its early vegetative growth develops into fruit production, it will ripen and swell fruit, grow new leaves, flowers, shoots and roots for as long as conditions permit. Each one of these stages of growth and development requires a complex mixture of nutrients, minerals, major and minor trace elements, in sufficient quantities, almost until the end of the season.

In a well made bed there will be enough phosphate for most of the season and sufficient nitrogen (N) and potash (K) until the first fruits start to swell. When the bottom truss is swelling, the second well set and the third in flower, it is time to start feeding. This may seem a little early but organic feeds need time to become soluble and available to the plant, unlike the rapid release of chemical feeds.

Nitrogen (N)

Over the months, tomatoes need a lot of N if small, pale-leafed impoverished plants, with their little pale green fruits that ripen to a deep red, are to be avoided. On the other hand, it is all too easy to feed them far too much N. Plants with huge dark green leaves, a thick stem, large shoots, long thin trusses standing at 1 o'clock to the stem, aborting or not setting their flowers, with the bottom trusses carrying large unwilling to ripen fruit, have had far too much available N. Not only is the plant unbalanced but is very prone to disease and insect pests find the soft sap all too enticing.

Potash (K)

Potash, needed for the overall health of the plant, is particularly important in the production of flowers and fruit. Here its main function appears to be that of a regulator for many of the metabolic processes in the cells as they make their proteins.

Research at H R I has shown that once a tomato plant reaches the fruiting stage it needs twice as much potash as nitrogen to produce top quality fruit than is needed to simply grow the plant itself. When yellowing margins appear, which spread and go brown, on the upper and middle leaves, they are usually followed by the uneven ripening and 'boxy' or hollow fruit that go with these symptoms of potash deficiency.

Phosphorus (P)

Phosphorus is well known as the nutrient needed for good root growth, and plant establishment. But it is also an essential element needed for the very life processes of the entire crop, the number of flowers and speed of development and maturity depends upon its availability.

Phosphorus, found in the soil combined with other elements as a phosphate, is highly insoluble. A soil may contain ample phosphate, but only small amounts are available to plants at any one time. This is directly connected to the pH of the soil, so the aim is to keep it around pH 6.0-6.7, too acid or too alkali and the phosphate forms insoluble compounds and becomes 'locked up'. When this happens and a deficiency occurs, the leaves stay small and are grey/green/violet/purple in colour. The leaves hang downwards and fall off with age, stems stay thin and the plant is slow to develop in every way.

Fortunately, although plants need a continuous supply of P it is only in small amounts and most lively organic soils with the right pH will supply this.

Balanced feeding

The aim when feeding tomato plants is to grow a well balanced productive plant. So, throughout the season, feed on a little and often basis, supplying twice as much potash as nitrogen for most of the season. Liquid tomato feeds both organic and inorganic are readily available and should be diluted according to the maker's instructions, as should the occasional feed of liquid seaweed which acts as a bio-stimulant and adds minerals and trace elements to the soil. Dilute liquid feeds may be given with every watering or every 2 or 3 days, depending on the speed of growth and the demands of the plant. The response is far quicker than that from solid feeds. Where dry organic or inorganic fertilisers are used it is even more important to follow the maker's instructions to eliminate the risk of burning the roots.

Traditionally home-grown tomatoes have been fed with home-made top dressings. These are many and varied often using readily available local materials, and they still have their devotees who each year grow some fine delicious fruit by regularly top dressing the beds. Top dressings and mulches should be applied to a moist soil and watered in; they are slower in action and often weaker in strength, so they need applying on a regular basis. The quantity in a handful varies but is not critical, but what is important is to wear a rubber glove when handling animal wastes to reduce the risk of infecting any cuts or grazes on the hands. The following quantities are fully variable but may be a useful guide.

Home made top dressings	Per plant	Applied every
Worm compost	Large handful	8–10 days
Old high potash manure		
from a covered stack	2 handfuls	8-10 days
Well made compost	2 handfuls	8-10 days
Wilted chopped comfrey	4 handfuls	10–12 days
Liquid comfrey	15 to 1 or 20 to 1	3 days
Pigeon Dip (Bag of)	20 to 1 to 30 to 1	When N
Poultry Dip (droppings	" "	boost
Sheep Dip (hung in)	" "	needed
Manure Dip (barrel)	" "	"

Comfrey

For a tomato grower Russian Comfrey, *Symphytum x uplandicum,* is the gardener's richest source of home grown potash. Its long roots penetrate deep into the subsoil and bring up the potassium, phosphorus, calcium and trace elements found in its leaves. The name Russian was given to the plant as it originally came from the gardens of the Palace of St. Petersburg, Russia. Henry Doubleday, a smallholder of Coggeshall, Essex, introduced the plant to this country and when Lawrence D. Hills founded the HDRA in 1954 (it did not become a registered charity until 1958) he named the association the Henry Doubleday Research Association in honour of the work done on comfrey by Henry Doubleday.

Bocking 14

Lawrence Hills set out to collect, grow, trial and analyse the different strains of comfrey being cultivated for animal feed. In the early years of the HDRA, Lawrence and his small team of researchers numbered the different strains and named them after the address of the research station, 'Bocking' Essex. The different strains were grown, evaluated and analysed. Eventually, after years of research, number 14 proved to be the most beneficial strain for all round garden use. Now whenever comfrey is mentioned it is invariably the 'Bocking 14 strain' that is being referred to.

It was in 1973 that Lawrence Hills suggested covering comfrey leaves with water to ferment for 3–4 weeks. The result was a useful tomato feed, but we all agreed with Lawrence that it 'stank to high heaven', as the proteins started to break down, releasing their strong odour into the water.

The most concentrated and least aromatic way of making liquid comfrey is to use the natural ability of the leaf to decompose rapidly. Wilted comfrey is packed into a barrel or container, preferably fitted with a tap; a board and weight are placed on top and the barrel covered with a lid or sheet to keep the rain out. Around 14-21 days later a thick dark brown liquid can be drawn off and may be accurately diluted to meet the needs of the plant.

I have been making liquid comfrey this way for over 25 years and find it an excellent tomato feed. However, being entirely natural, it varies from season to season and batch to batch. The strength of the liquid depends on the weather, date of cut, size and water content of the leaf and the health of the comfrey bed itself. Over the years, without the benefit of sample analysis, I have found that when the liquid comfrey is thick and black it would seem to be at its strongest and use it diluted with 20 parts of water to 1 part comfrey or for a weaker feed 30 to 1. During a very wet spell when the comfrey leaves are soft and sappy the resultant exudant is thinner and brown in colour, in this state a dilution of 15 to 1 or even 10 to 1 gives better results. This is very much a frequently changing ratio depending on the plant's needs, but as a rule of thumb it works.

Comfrey liquid made early in the season tends to contain a more balanced mixture of nutrients sometimes with a higher nitrogen content in it. By September, the potash content of a comfrey leaf is at its highest and an exudant made from these leaves will make a liquid feed which usually contains around twice as much potash as nitrogen, making it an ideal feed for a tomato plant. And all it costs is a little effort.

Comfrey leaf mould

Comfrey cut in September when the potassium levels are high and mixed with 2- or 3-year-old leaf mould makes a nutritious, free rooting propagating medium. Place a 3-4 "/8-10cm layer of leaf mould in the bottom of a large strong plastic bag, add a layer of wilted comfrey of the same thickness, then a layer of leaf mould, another layer of comfrey and so on until the bag is almost full, tie up the top to keep the rain out, make a few holes in the bag with a fork to allow air in and store in dry shade or under cover until spring.

In spring tip the mixture onto a clean surface and, apart from the odd stem, the comfrey will have been absorbed by the leaves. Riddle the mixture and add sharp sand as required, check the pH; it is usually around pH6.5, but if the leaves have been acidic add the required amount of Dolomite Limestone as indicated by the test and the compost is ready. Over the past 20 years that I have been making comfrey leaf mould it has produced some fine plants with only slight seasonal variations. If the leaf mould is 'clean' and the comfrey cut clear of the soil the compost is usually free from soil-borne pathogens and normally contains enough nutrients for the propagating stage of tomato plants up to the planting out stage.

Tomato plants grown in 2 gallon/10 litre pots of comfrey leaf mould thrive for the first 2-3 trusses, but it would be unrealistic to expect even the best compost to sustain plants up to the 6-8 truss range without further nutrients in the pot. There are many compost based dressings now available, both organic and inorganic, and when a full crop is being grown, add a suitable base at the recommended level into the compost at the mixing stage. Where top dressings are going to be used, initially leave sufficient room in the top of the pot to accommodate the first dressings, as the season goes on the compost will oxidise, settle and sink leaving room for later top dressings. Comfrey leaf mould is an entirely natural basic material and once again all it costs is a little effort.

Summary

- Feed little and often with top dressings.
- Use a balanced feed for vegetative growth.
- Use a high potash feed for fruit quality.
- Potash to nitrogen ratio 2 to 1 once two trusses are half swollen.
- Prepare for feeding in advance with stocks of suitable top dressings.
- Prepare in advance supply's of comfrey liquid and/or manure 'dip'.

17 Mulching the borders

Traditionally tomato beds have been mulched about the time the first tomatoes have started to ripen; in the days before ready-made tomato feeds mulching was a vital part of crop management, and it is no less important today. A well made tomato bed with its bulky organic matter content is a living mass of actively productive bacteria and fungi busy converting the nutrients into plant food.

The border soil has 3 main groups of microflora, bacteria, actinomycetes and fungi. The actual numbers and balance or these in any particular tomato bed varies with the soil type, but whatever the basic soil type, they do have one thing in common, a really active bed settles and sinks as the season goes on.

Water pressure

Soil, watered with a fast flowing hosepipe, or a jet of water from a watering can held at waist height, is subject to considerable water pressure. Over the weeks this has a compacting effect on the surface of the bed, not only by direct pressure but by washing the finer particles down into the pore spaces of the soil.

With the anticipation of those first ripe tomatoes in mind, the level and condition of the border soil often goes unnoticed, when really this is the very time to be taking action. A close look reveals that the settling and oxidation of the soil may have reduced the surface level by up to 1"/2.5 cm. As repeated jets of water wash away the surface compost from the original root ball, the tops of the root system are revealed, exposing them to both heat and light.

Individual fragile feeding root hairs only live for about 3 days before dying and being replaced. When tomato roots are too hot the growth rate slows down, exposure to heat well and truly cooks any fine surface root hairs before they have had time to contribute to the plant. If this heat should coincide with naturally occurring root death it adds to the check and the plant suffers.

Naturally occurring root death

Tomato plants often appear to slow down or stop growing when the first tomatoes are ripe. This is the time when the plant is under maximum stress, it is ripening tomatoes, maturing green fruit, swelling fruit on another 5 or 6 trusses, making viable pollen and fertilising flowers while growing new leaves and trusses. All this contributes to slowing the rate of growth, but the real change is going on underground. Scientists at the Glasshouse Crops Research Institute found that by the time a tomato plant is around 120 to 130 days old the plant suffers from naturally occurring root death which effectively checks the plant.

When tomato plants are only being grown up to 5 to 6 trusses and this check coincides with 'stopping' the plant so often it goes unnoticed, but for a long season's crop the earlier the plants start to make new roots the better. There may be little evidence of returning to the old speed of growth until those bottom trusses have been picked and the plant is under less pressure. In severe cases of stress the plants may fail to set the next truss or two until the root/shoot balance has been restored.

Surface water evaporation

A proportion of the water and liquid feed applied to the exposed surface of a tomato bed is lost to the plant through the cooling effect of evaporation. The rate of evaporation is controlled by the proportion of the energy absorbed by the soil which is used for evaporating water and the speed of the removal of the water vapour from the spot where it is produced. By midseason the buoyant atmosphere that is created by encouraging warm air movement to aid ripening quickly removes water vapour. Hot weather speeds up the cycle and the plants need more water just to replace surface loss.

Mulching

The answer to these various problems is to mulch. Mulches spread over the border surface as part of the feeding programme also work as a temperature equaliser. In spring a soil temperature of 57°-58°F/14.5°C is needed before tomato plants will even root. With the soil at between 60°-70°F/16°-22°C, the root system quickly expands and thrives. Once the border reaches this degree of warmth, it can rise rapidly in hot weather. When it is hot a mulch acts as a blanket and reduces the wide temperature variations from dawn to noon which can be as much as 20°F/12°C instead of the more beneficial rise and fall of 4-7°F/ 2°-3°C.

By the time the first fruit are ripe the bottom leaves will have been removed up to the first truss, making spreading easier. After evenly watering the surface of the whole bed, about 1"/2.5cm of mulch is applied. When the first truss is fully picked, another 1-2"/2.5-5cm is placed on top of this. If the first mulch is of worm compost (made by feeding vegetable waste to a colony of worms in a dustbin-sized conatiner) it provides a nutritional boost just at the right time, but well made compost and old manure also provide nutrients and good cover, these may also be used for the second layer.

Chopped wilted comfrey is also an excellent mulch. This needs regular topping up in order to maintain the volume, as it decomposes rapidly. Under the shelter of the mulch the renewed and reinvigorated soil microflora multiply rapidly, improving the soil surface permeability and, taking advantage of the moist conditions and the added nutrients, surface roots start to grow again.

Summary

• Water and or feed the bed before mulching.
• Have a supply of mulching materials ready to apply.
• Top up layer of mulch when it is needed.

18 Tomato ripening

The whole culture of the tomato plant is directed to one end: ripe fruit. Succulent sun ripened, home grown tomatoes have that distinctive taste only this superb fruit can provide. During the last fortnight before ripening, when the fruit is at the mature green stage, the process seems to be tantalisingly slow, especially in cold greenhouses.

The process of ripening is triggered and governed by temperature with the optimum at 68°F/20°C but the range 65-75°F/18.5-24°C seems to suit most modern cultivars. Given good growing conditions, the genetic clock built into most modern cultivars ticks off some 55 days between the fertilised flower and the ripe fruit. A few varieties are earlier, while some beefsteaks take almost twice as long.

Internal changes

Before even a hint of colour is seen, dramatic changes must take place within mature green fruit to prepare for pigment production. As the chlorophyll starts to break down, so does the starch into a little glucose and, hopefully, a lot of fructose which is increased by sunshine: hence the adage, 'more sun more sugar'. At the same time, the ratio of citric acid to malic acid is becoming more citric, which aids the flavour. This depends on the potash levels in the plant; as the old saying goes, ' more potash more acid'.

Ethylene production gets under way. This powerful hormone stimulates ripening with as little as one part per million. The cell walls start to soften under the influence of the production of polygalacturonase, and that wonderful tomato flavour and aroma develops as the volatiles form.

Pigment formation

Red tomatoes are coloured by a mixture of two pigments, beta carotene and lycopene, with the actual ratio varying with the variety. Dark green mature fruit usually have more lycopene in their makeup. This is the pigment that determines the final degree of redness. Lycopene is slow to form at low temperatures, hence the slow ripening in cold weather. At the other end of the scale, should the fruit mature in a heatwave, at temperatures over 86°F/30°C lycopene production stops altogether.

Beta carotene is the dominant pigment of those cultivars that are consistently light green prior to ripening. These fruits colour evenly to a mid orange when ripe and continue to ripen at high temperatures.

The vitamin A content of tomatoes is directly linked to the amount of carotene present and it is considerably higher in the bright orange coloured cultivars.

Carotene also affects the difference in flavour and aroma between red and orange coloured tomatoes; this is due to the different range of volatile compounds produced by cultivars with a high beta carotene content, compared with the volatiles produced by varieties with a high lycopene content.

Volatiles

That unique, herby, musky, distinctive smell given off by ripe tomatoes that starts the taste buds tingling comes from the volatile compounds in the fruit which increase with ripening and are important to the overall flavour.

This distinctive tomato aroma comes from a wide mixture of some of the over 200 complex chemical compounds so powerful that they are easily recognised even when they comprise only up to 1% of the fruit. The aroma is more pronounced in some varieties than others and differs over the season but there is no mistaking its presence. It is not possible to create any specific mixture of volatiles but plants which grow with a full complement of nutrients, minerals and trace elements usually advertise their presence by their appealing distinctive aroma.

Once bottom trusses are mature enough to ripen, temperature management is based on trying to maintain a daytime temperature of 68°F/20°C, plus and a night-time one of 60°F/15.6°C. At this stage of plant development the bottom leaves will have been removed and the warm air should be able to flow freely among the bottom trusses to start the ripening process.

Summary

- Fruit needs to be old enough for ripening to start.
- Temperature is the trigger it needs to be at least 68°F/20°C.
- Deep red tomatoes contain more lycopene than beta-carotene.
- Orange red tomatoes contain more beta-carotene than lycopene.
- Bright orange tomatoes contain much more pro-vitamin A than others.
- Leaves removed up to first truss speed ripening.

19 Stopping the plants

Knowing exactly when to stop tomato plants depends on several factors. When the greenhouse is already booked for a following crop, a definite date has to be set. Eight weeks before the crop is to be removed the plants are 'stopped' by pinching out the growing point. Thus plants stopped in mid July will be almost picked clean by mid September. When a crop goes on until the end of October or into November with falling light and temperature levels, the timing alters. Under late autumn conditions, the 8 weeks flower to fruit timing extends to 10 or even 12 weeks. So for a mid-November finish stop the plants in mid-August.

In a sunny greenhouse it may be worthwhile aiming for an extra late truss by stopping the plant 2 leaves above any truss with the flowers just showing colour in mid–August. If a month later the flowers have set well and the fruit is swelling steadily, leave it on. If the weather is poor and the truss does not develop well, then cut it off cleanly allowing the lower trusses to develop fully.

Apical dominance

When a tomato plant is stopped any remaining leaf axil shoots grow apace, so the plants still need regular trimming.

Older leaves need to be removed but still only up to a ripening truss as the young leaves are now photosynthesising less each day as the light levels fall. With falling light and lowering temperatures, watering is gradually reduced. Octobers seem to be getting warmer and some years St Luke's summer, around October 18th, arrives with a spell of warm sunny weather when the plants will need more water, but not too much or split fruit will result. Over the last few weeks of the season watering is reduced and for the last fortnight the beds are allowed to dry out.

Summary

• Time plant stopping to allow most of the fruit to swell and ripen naturally on the plant.
• Early stopping: 8 weeks is sufficient time.
• Late stopping: allow up to 12 weeks.
• Continue to remove new shoot growth.
• Leaf up to a ripening truss.
• Stopped plants need progressively less water.

20 Outdoor tomatoes

Looking through the list of tomatoes there are varieties noted as being suitable for outdoor growing and there are many more listed for growing in cold greenhouses or outdoors in a sunny spot. This is a sign of the changing climate: just a few years ago, outdoor tomatoes were regarded as a southern crop, from which only gardeners South of a line between Bristol and the Wash could expect a reasonable result, with the rest of us making the most of what was often an 11- or 12-week season. While it is still true that the further North anyone lives the number of weeks when the mean temperature is conducive to outdoor tomato growing is less than South of the Bristol/Wash line, things are changing.

Gardening here at a latitude of 53° N, we have been able to grow up to 7 trusses outdoors in 3 of the last 5 years, with gardeners in parts of Scotland getting good results from fast maturing dwarf varieties.

Potato or late blight (*Phytophthora infestans*)

Along with the extended frost-free season have come frequent periods of warm damp weather encouraging ever more potato blight. Areas of the country which in the past have only had the occasional attack are now finding that blight is a regular occurrence and that growing blight susceptible varieties is a waste of time.

Fortunately the scientists and plant breeders have come up with some first-class blight-tolerant and blight-resistant new varieties, which, together with other built in disease resistances, are producing some lovely outdoor fruit across the country.

Timing

To get the very best from outdoor tomatoes it is imperative that as much growth as possible is made before the plants go out and this is even more important in northern areas. Choose from the wide selection of suitable varieties, then sow early. Depending on the facilities available, the first day of spring (March 21st) is not too soon to sow the varieties chosen for basket, pot and cloche culture. Varieties destined for the open ground may benefit from sowing a fortnight later.

Basket and pot culture

Plants grow rapidly in April and with so many demands on space care must be taken not to allow overcrowding to 'draw' them. By mid–May aim to have a very robust plant with the first flowers opening, ready for their final potting. Vigorous trailing dwarfs can go straight into baskets filled with potting compost and hung wherever there is space; when the time comes to put them outside in late May or early June one plant will have filled the basket and should be smothered with small fruit starting to swell.

With a whole season's growth to contain, choose a large pot for dwarf and medium dwarf varieties and for the larger types try a 3-gallon/13.6 ltr bucket with drainage holes drilled in the bottom of the sides, rather than holes in the base of a flat bottomed bucket which may not drain when placed on a flat surface.

Potting

Fill the pot two thirds full with potting compost to allow for later top dressings and put in the plant. Tie initially to a 2-3 foot/60-90cm cane. In all but the most sheltered districts it is far too early to place them in their final growing positions so keep the plants in a light frost-free place. However, these March sown plants ensure that a big plant is ready to go out as soon as conditions allow.

Finding space

If the greenhouse is bursting at the seams, or one is not available, make good use of the fact that during late spring light values are some of the highest of the year. This means that temporary frost protection can be provided for 10 to 20 days, in porches, on spare bedroom window sills, garage windows, or a sheltered corner with a lean-to plastic sheet. The plants may 'draw' a little but they should not suffer unduly. The flowers will be opening so at this stage hand spray with water and tap the canes to set those valuable early fruits.

During the last few days before placing the pots in their final positions, gradually harden the plants off. When the danger of frost has passed place the pots in their final growing positions. Dwarfs and bushes may need additional support or be left to sprawl over the sides of the pots. Replace the small cane in the pots of cordons with a 6 foot/1.8metre one and tie the cane top to a horizontal wire or other support. Pots standing in shallow saucers reabsorb nutrients drained into them without becoming waterlogged.

Site

Very good crops can be grown using the residual heat from a South-facing wall. If the wall is white, so much the better, but any sheltered spot that may be sunny for most of the day can be used. Shelter for outdoor tomatoes means protection from wind and wind-driven rain. It really is a case of trial and error but anywhere in a wind tunnel or in the path of a small whirlwind is best avoided.

Simple propagation

Where a greenhouse is unavailable for propagation and plants are going into the open ground, useful crops can still be grown as late spring is the time for windowsill box-grown plants. Choose a dwarf, fast growing, quick cropping variety such as Red Alert or Totem; grow from seed, or buy the seedlings as soon as they are available locally, plant 4 into a polystyrene wallpaper trough available from DIY outlets and grow on a sunny windowsill. During May squat broad plants are produced; these will gently tease apart with a mass of roots just bursting to grow away and establish the plant.

This large comparatively shallow root ball is a big advantage in the open ground. The soil temperature of 57°F/14.5°C, needed for root growth rarely reaches down below the top 3 "/7.5cm of soil in early June. When the soil is only surface warm one way to get the plant off to a quick start is to take out a larger planting hole than is needed and back fill with the warm surface soil, it will cool a little, but even an odd degree helps.

Outdoor borders

Depending on the district, outdoor tomatoes all too often face a cold period in June, a wet one in July, and a soggy spell in August interspaced with dry sunny periods. In order to balance out the soil moisture as much as possible and maintain its strength to produce high quality fruit, both excellent drainage and good water retention are needed.

The ground for a row of plants, preferably sheltered and South-facing, may be prepared as a trench. The subsoil is 'cracked up' with a fork and a dressing of well made compost and or well rotted manure is worked into the top spit. High nitrogen manures are best avoided as combined with wet warm weather they will produce far too much soft, disease-prone growth at the expense of fruit. A fortnight before planting work in 8 oz/225g of organic base dressing per yard/metre of trench. If the border is covered with a strip of black polythene, cloches or frames this will help to warm up the soil. The plants should be thoroughly hardened off, and as soon as frost danger is past it is time to plant.

Planting and cloches

Put in a stout cane or stake, trowel out a hole large enough and plant with the root ball just slightly above the soil level, taking care to keep the garden soil off the root ball to help keep the legions of lurking pathogens away from the stem for as long as possible, a loop of twine and a pint of water each and the job is done.

Although not as popular as they were in the past, cloches can be very useful when growing dwarf varieties. Cloches placed over the prepared tomato beds 2-3 weeks before planting warm up the soil and plants can go out 10-14 days earlier than into the open border. Dwarf varieties grown this way need no staking, tying or trimming, being easy to grow, they have much to recommend them.

Top dressing

A month after planting, hopefully the soil is warm enough to mulch with chopped wilted comfrey, well rotted manure, or best of all worm compost. Pot grown plants should be top dressed in this way at this time, and again a month later. The value of top dressing and mulching is seen after a prolonged wet spell when the rain has leached out the soluble nutrients particularly from soils that are only fed with liquid feeds. On these soils soft growth follows, with large flavour-lacking fruit produced. Under these conditions a couple of potash rich liquid feeds will help to keep the plant in balance.

Watering

Basket and pot grown plants need watering every day in dry weather and will benefit from a liquid feed a couple of times a week. It is only in drought conditions or on very fast draining soils that border plants need watering; for all others a well prepared bed will usually hold enough water for the plants' needs. Cordon varieties need to be regularly tied in and are stopped in early August South of the Bristol/Wash line and towards the end of July North of it.

Season's end

For most of the summer outdoor tomatoes require little attention. Cordons need their side shoots pinching out and any excessive growth removing from prolific bush types, only dead or diseased leaves are removed and the plants kept weed free.

As sprawling bush varieties start to ripen their fruit, they do benefit from a layer of straw placed underneath them, this increases the air flow which helps to keep them healthy and lifts the ripening fruit clear of the soil keeping it free from soil splashes.

In recent years early autumn has been frost free and mild, with fruit ripening well into mid-October; the day comes however when wind and rain signal the end of the season. Mature fruit are picked off and ripened indoors and when dwarf varieties are still carrying a lot of fruit the whole plant can be cut off and hung upside down in a warm airy place to finish ripening.

After a good outdoor crop it's tempting to say, 'Who needs a greenhouse?'. After a bad one it's easy to say, 'Never again'. My first crop of outdoor tomatoes, 5 trusses of Moneymaker, are recorded on a black and white photo taken in 1956. Since then, on balance, I would say that they have been definitely worthwhile. With such a range of new, heavy cropping, disease resistant, delicious varieties available, anyone with a sunny spot, be it window sill, balcony, patio, garage side or open garden can now have the pleasure of growing and enjoying their own plant fresh, warm, succulent tomatoes.

Summary

- Choose varieties to suit method of growing, site and length of season.
- Grow the biggest plants propagating facilities allow.
- Providing early protection for pot grown plants increases yields.
- Do not over feed soil grown plants as lush plants =less tasty fruit.
- If possible protect from damaging winds.
- Temporarily cover with fleece when early frost threatens.
- Pick off fruit, or cut and hang in warmth before severe frost arrives.

21 Disorders, pests and diseases

The whole aim and object of home-grown tomatoes is to grow them as naturally as possible; strong healthy plants are able to cope with pests and diseases far better than weak unhealthy ones. Positive healthy culture should be the aim and with so many disease-resistant varieties now available sound crops can be grown with confidence. However it would be unrealistic to expect that everything in the garden will always be lovely and at times problems do occur, some of the most common ones are:

Physiological disorders

Magnesium deficiency

Magnesium is the central atom of chlorophyll molecules, so deficiency quickly shows when the area between the leaf veins goes yellow, leaving the veins green; when severe, leaves go completely yellow and die. Often seen on older leaves at the time the fruit starts to ripen, then spreads up the plant. A 2% solution of Epsom salts (magnesium sulphate) applied as a foliar spray once a week, or better still a spray of liquid seaweed of the manufacturer's rate of dilution will keep the leaves green.

Blossom end rot

A sunken brown or black area on the base of a swelling tomato is the first sign of blossom end rot, which is caused by a shortage of available calcium. One of the forms of calcium in the fruit is calcium pectate which acts as a 'glue' in the development of the middle lamella, strengthening and sticking together the cell walls; when the amount of calcium falls below 0.5% the cells become unstuck. Usually there is suffi-cient calcium in the soil which is transported up to the fruit on the water stream; if there is a shortage of water there may be a shortage of calcium.

Good tomato flavour is the result of strong soil solutions, blossom end rot is a consequence of too strong a soil solution. It is all a question of balance: sufficient feed for good flavour, enough water to transport the nutrients. Growbags with their limited capacity quickly dry out and to prevent blossom end rot may need watering several times a day during hot weather.

Pests

White fly (*Trialeurodes vaporariorum*)

Severe infestations of this sap sucking insect can weaken the plant and the sooty moulds attracted to the 'honeydew' excreted by them disfigure the fruit. When white fly appear late in the season, a few yellow sticky traps hung among the plants will help to control them. When white fly are seen early in the season, contact one of the suppliers of the parasitic wasp *Encasia formosa* who will supply details. Timing and temperature are important in getting the host/parasite ratio right for successful control. Many gardeners find that planting French marigolds among tomatoes helps to deter an attack of white fly.

Red spider mite (*Tetranchus urticae*)

Probably the most insidious glasshouse pest is red spider mite. It arrives unnoticed, usually on bought in plants, multiplies rapidly, until one day the plants appear dry and crusty, the leaves covered with white pinprick holes often with a yellow mottled appearance. Seen through a strong hand lens the mites are an ugly pest, red, light brown, or greyish green in colour; in their thousands they can suck out the very life force from a tomato plant. Mites do not like water and regular top spraying can retard small colonies but the only real answer is biological control; when mites are seen contact one of the suppliers of the predator *Phytosiulus persimilis* for full details and supplies.

Diseases

Botrytis cinerea (Grey mould)

The most ubiquitous of pathogens affecting both hot, cold and outdoor crops of tomatoes is *Botrytis cinerea* – grey mould. As the hosts for this fungus are to be found among most types of plants, given the right conditions it is almost unavoidable.

At a temperature of 68°F/20°C, and above, when the relative humidity is 90 to 100%, it only takes 5 hours for botrytis spores to germinate. The light to mid brown lesions on stems and leaves turn grey as the spores develop and become covered in the furry looking grey mould that releases thousands of spores into the air when touched.

On the fruit the spores cause the small white circles which when numerous disfigure the skin; when the calyx is attacked at the point where it joins the tomato wet rot forms and the fruit drops off.

The only effective answer to botrytis is prevention. Keeping the greenhouse clear of all plant debris helps, as does allowing as much ventilation as the weather conditions permit. One way to tackle stem lesion is by surgery with a very small knife; success depends on how much has to be cut away and how quickly the wound can be dusted with a coat of flowers of sulphur.

Cladosporium fulvum **(Leaf mould)**

Warm moist stagnant air helps the spores of *Cladosporium* to spread. This disease is seen as yellow or bright orange leaf spots with brown felty undersides which coalesce to fill and kill the leaf. Many of the older varieties are susceptible to the disease and to restrict it are better planted wider apart and given as much air as conditions allow. New varieties with C. 1 to 5 or C abcde after their name carry resistance to the disease.

Verticillium **and** *Fusarium* **wilt**

These soil borne fungus diseases are still present in many soils. A wide selection of varieties marked V or F are now available with inbuilt resistance to them.

Virus

Tobacco Mosaic Virus

There are several viruses that attack tomato plants, but tobacco, or as it is often called, tomato mosaic virus, is the commonest. The virus appears as twisted, deformed foliage, with light or dark green or yellow mottling, new leaves may be fern-like in structure, and the plant becomes stunted. The virus is very easily spread by insects and by the hands and knives when tending the plant. Infected plants should be carefully removed and burnt and hands and tools disinfected. Fortunately there are now many varieties with TMV after their names denoting resistance to what was once a widespread scourge.

22 Types of tomato

Tomato plants have different types of growth; these are:

Dwarf Rarely more than 15"/45cm tall, self supporting, sometimes more than one stem, does not need trimming or pruning, e.g. Tiny Tim, Totem.
Trailing Dwarf Dwarf form having several stems suitable for growing in and trailing over baskets and troughs, e.g. Red Alert, Tumbler.
Determinate, Det., Bush This type varies in height from 2-4 foot/60-120cm. When they reach their natural varietal height, determinate types stop growing upwards and produce trusses, flowers and fruit on their several stems which there is no need to

prune. Once ripening starts, some varieties will ripen all their fruit within a couple of weeks or so. Smaller varieties, e.g. Histon Early, may be left to sprawl without further attention; larger ones, e.g. Roma, benefit from some support, such as a stake and twine, a circle of netting, or short twiggy sticks.

Semi-determinate Some varieties are between the determinate and the indeterminate and may grow either way; if in doubt support 2 shoots and allow to flower, set and swell some fruit, then, if excessive growth appears, prune to a manageable shape which allows for good air circulation.

Indeterminate or cordon These types will continue to grow upwards for as long as the growing point remains intact. When the temperature is warm enough and the light strong enough most glasshouse cultivars will grow 2 new leaves and 1 new truss every week, until they are 'stopped' or the good growing conditions cease. In practice most greenhouse tomatoes are 'stopped' by having their growing point pinched out when they have filled their allotted space.

Definitions Descriptions Abbreviations

Open pollinated These are self-fertile plants replicating the same characteristics for generation after generation, e.g. Ailsa Craig.

F1 This denotes 'First filial generation', a hybrid resulting from a first cross of two distinct parental lines, e.g. Shirley F1.

Cultivar The name of a variety of a plant, e.g. Cultivar Brandywine.

Beefsteak The name Beefsteak is used to describe a variety capable of reaching a very large size, often over 1lb/500g in weight, that has solid or meaty flesh with few seeds and little gel.

Heirloom An old variety that has been handed down through the years in a locality and sometimes through many generations of the same family.

Bilocular When a bilocular tomato is cut across its axis it reveals 2 locules or chambers containing the seeds and their surrounding jelly (see opposite).

Trilocular Containing 3 locules (see opposite).

Multilocular Containing several locules (see opposite).

Greenback Varieties genetically susceptible to greenback have dark green shoulders on the unripe fruit which sometimes stay green or yellowish green when the rest of the fruit turns red; greenback is a physiological condition. It is most noticeable when those susceptible are grown near the glass or ripened in very high air temperatures. Most modern F1 varieties are greenback resistant.

Mature Green A description of a tomato that has reached its full size and is mature enough for the ripening process to start.

Disease resistance symbols

The letters placed behind tomato variety names in catalogues and on seed packets to indicate that particular variety's resistance to specific diseases.

A=*Alternaria.* Soil-borne fungus causing fruit spot and rot.

V = *Verticillium dahlia* and *Verticillium alboatrum.* Soil borne fungus disease.

F = *Fusarium oxysporum lycopersici.* Formerly strain 1 now strain 0.

F2 = *Fusarium oxysporum lycopersici.* Strain 0 formerly strain 2.

Bilocular: Ailsa Craig

Trilocular: Green Zebra

Multilocular: Black Crimea

Fr = *Fusarium oxysporum radis- lycopersici.* Fusarium is a soil borne fungus disease

P = *Pyrenochaeta lycopersici.* Fungus causing corky root disease and root rot.

N = *Meloidogyne* species. Gall nematodes cause root gall.

M = *Phytophthora infestans.* Air borne fungus causing late or potato blight.

C = *Cladosporium fulvum.* Races 1 to 5 or a, to e. Fungus causing tomato leaf mould.

S = *Stemphylium* species. Causes leaf spot.(Sometimes listed as St)

T or TMV = Tomato Mosaic Virus. Causes mottled foliage and degenerative disease.

Pt = *Pseudomonas syringae* pv.tomato. Causes speckles or spots on fruit.

Descriptions

In the following descriptions of tomato varieties all Determinate types will be referred to as **Bush** and all Indeterminate types referred to as **Cordon** as these are the terms used by most people.

The numbers given after the tomato types are an approximate guide to the number of days it takes for the first fruit to ripen after the plant is planted out, given normal growing conditions. Where no numbers are quoted, usually the small fruited varieties are the earliest to ripen, then the standard sizes with the beefsteaks taking the longest. Under adverse weather conditions add on the number of poor days to the time estimated to ripen. Late planted, late varieties in a poor summer can take up to 14 weeks to ripen.

TOMATO VARIETIES

Fablonelystnyi

DWARF AND TRAILING DWARF VARIETIES

Balconi Red & Balconi Yellow

Type Bush, does well in baskets or containers.
Site Outdoors, preferably in a sunny spot.
Fruit Small, delicious, bright red or yellow.
Plus points Attractive, edible, garden feature, very productive and sweet.

Here are two of a kind, offering masses of small tomatoes on sturdy, bushy compact plants specially bred for container, basket or trough growing.

    The plants are capable of carrying a very heavy crop which pulls the trusses over so are ideally suited to hanging baskets where the lax habit of the plants allows them to hang unrestricted over the side. Balconi Red crops well but of the two Balconi Yellow is the heavier cropper with clusters of bright yellow small deliciously sweet fruit ripening over a long season. Baskets of Balconi Red and Balconi Yellow are attractive, colourful garden features in their own right as well as being prolific croppers over many weeks.

Garden Pearl or Gartenperle

Type Dwarf trailing bush.
Site Outdoor border. Baskets or tubs.
Fruit Small, pink to rosy-red.
Plus points Ideal for baskets or troughs.

This is a fine variety for containers in sunny spots and is equally at home in a hanging basket where the trailing habit of the trusses hang down in profusion. Planted out in a sunny border, one plant can cover over 1 square yard/metre with a mound of short trusses covered with reddish pink cherry-sized tomatoes. In the open border when the plants are left to sprawl where they will, the huge crop of small tomatoes do get rain splashed unless supported by a layer of straw or allowed to grow over a layer of twiggy sticks saved from the winter pruning.

Red Alert

Type Dwarf bush.
Site Inside or outdoors.
Fruit Small, up to 2oz/56g, sweet sharp, lingering taste.
Plus point Dual purpose, excellent inside or outdoors.

Red Alert seed should be on every tomato grower's shopping list. It is best known as a very early outdoor dwarf bush. Planted in a container or basket and put out after the last frost, it quickly turns flowers into delicious fruit. Expect each plant to yield about 5-6lb/2.5kg of glorious cardinal scarlet fruit.

Recent months have seen fuel costs rise yet again making every bit of heated glasshouse space even more valuable. A few seeds of Red Alert sown as early as possible will provide plants for hanging baskets or containers which can be easily moved around the greenhouse or into any available sheltered space. Used this way, Red Alert can provide the earliest possible ripe fruit, weeks before the main crop is ready.

Tiny Tim

Type Tree tomato   55 days.
Site Pot or small trough. Inside or in a sunny outdoor spot.
Fruit Small bright red, ornamental and sweet.
Plus points Will grow in anything. Children love growing it.

The term 'tree tomato' was commonly used in the past to describe some of the dwarf tomato varieties and has recently come back into use with the promotion of the so called 'Tree tomato' *Tamarillo cyphomanora betacea*. Although a member of the Solanaceae family, the Tamarillo is not a tomato but a half hardy perennial Peruvian fruit, growing on a woody plant 6-10 feet/2-3 metres tall.

Originally 'tree tomato' was the name given to a tomato variety with a stem strong enough to be self-supporting and goes back to the days of 'Dwarf Champion' in 1865. When Dwarf Champion was crossed with Redskin one of the offspring was Windowbox which in turn was crossed with the wild tomato *Lycopersicon pimpinellifolium*. The result was Tiny Tim introduced in 1944. Try Tiny Tim in a pot or put 3 into a small trough and place into a sunny spot where they only need to be fed and watered. No more than 2 foot/60cm tall, the plant is self-supporting and becomes covered with ¾"/2 cm bright red fruit. A useful ornamental feature as well as providing sweet juicy fruit and a child's personal plant.

Totem

Type Neat compact dwarf.
Site Outdoors or sunny indoor window sill.
Fruit Glowing crimson, delectable flavour.
Plus points Heavy cropping container variety.

This variety, which originated in Russia, is ideal for container cultivation, or where space is limited. The dark green short leaves on a stocky stem make Totem a most attractive looking plant, which becomes covered with stubby trusses bearing small round glowing crimson fruit with a most appealing flavour.

Reaching a height of 18-30"/46-76cm tall, Totem is quite capable of producing up to 10lb/4kg of lovely tomatoes either in pots or in the open ground.

Tumbler

Type Trailing cascading foliage.
Site Outdoor or any cold structure.
Fruit Bright red bite sized sweet cherries.
Plus points Very versatile, tumbles anywhere.

The name says it all. Tumbler is one of the best trailing vines for container growing. One plant will fill a hanging basket, two plants in a trough will cascade down a wall providing a decorative edible curtain.

For many years I have found Tumbler to be the best variety for pannier growing. In a sunny spot a grow bag is hung across a narrow wall, fence, low branch or stout wire, creating a pannier with its contents equally in each half. A slit is made near the top of each side of the pannier and a Tumbler planted in each slit. Kept watered and fed, Tumbler will keep producing small sweet cherries throughout the summer and when grown under cover crops until late into the autumn.

Determinate or Bush Types

Bush Lilliput

Type Sprawling bush.
Site Outdoors or spacious cold greenhouse.
Fruit Golden, small, sweet and juicy.
Plus points Trouble free sprawler.

Bush Lilliput is one of those varieties that can be planted and left to their own devices until it is time to pick the fruit. Ideal for the busy gardener, this variety makes a sprawling mound of foliage with its trusses on top for easy picking and ripening well clear of the soil.

With a sweet juicy flavour it is a good variety for a garden with plenty of space but a gardener who is short of time.

Elios F1

Type Vigorous bush.
Site Outdoors.
Fruit Solid, thick walled prolific.
Plus points Superb kitchen fruit. V. F.N.A.St.

Tomatoes destined for sauces, soups, roasting and bottling need to have thick walled meaty flesh with a full bodied rich flavour. The lovely Elios fits this description admirably with the added attraction of having a high level of resistance against a wide range of soil borne pests and diseases. This vigorous bush variety produces a large crop of most attractive looking first class fruit on most soils. As with all sprawling varieties, a handful of straw under the plants will help to keep the fruit clean.

Histon Cropper

Type Sprawling dwarf.
Site Outdoors.
Fruit Small to medium with a good flavour.
Plus points Early outdoor, attention free, has good blight resistance.

Histon Cropper and Histon Early are a pair of bush tomatoes bred to provide early field or garden crops with a good flavour that require the minimum of attention. They are both good varieties with Early Histon ripening bright red fruits 10 to 14 days before its sister, but Histon Cropper with its good blight resistance may go on cropping until late summer. As with all sprawling varieties, a handful of straw under the plants will help to keep the fruit clean.

Elberta Girl

Type Vigorous bush.
Site Needs space either in a cold greenhouse or outdoors.
Fruit Beautiful, distinctive, striped semi plum shaped.
Plus points One of the loveliest tomatoes to look at and taste.

From the southern states of America comes this silvery leafed plant bearing truly beautiful tomatoes. When fully ripe they are a bright brick red with each fruit covered with random stripes of regal gold. With a round slight plum shape the profuse 3oz/85gm juicy fruit have a distinctive 'tang' all of their own.

Kotlas

Type Vigorous small bush.
Site Any sunny outdoor spot.
Fruit Bright red, sweet and juicy.
Plus points Highly portable. Blight tolerant.

Kotlas, sometimes listed as 'Sprint', is a little gem of a plant. Originally from Russia, it is another cold tolerant dwarf bush that is equally at home in a sunny border as it is in a pot. At first impression it appears to be a very slight little plant with its sparse small leaves and compact habit, but those leaves carry a high level of blight tolerance.

Kotlas is one of those varieties that can be potted into an 8 or 9"/20- 23cm pot or container and tied to a short cane, moved to wherever there is a sunny spot where it will yield a surprisingly good crop of small round 2oz/56gm tomatoes with very good keeping qualities and a rich full bodied flavour. A wonderful variety to use as an early season taster before the main crop starts to ripen.

Legend

Type Medium sized bush.
Site Cold house or outdoors.
Fruit Large and juicy with few seeds.
Plus points Exceptional blight tolerance.

Legend is the latest American raised blight tolerant tomato on offer. It is a vigorous, strong bush and planted outdoors with- stood an attack of late blight here in Yorkshire in 2005. The fruit come into the early ripening class and continued to ripen steadily on the plant until the end of September. Mainly round and solid with few seeds, Legend is nevertheless juicy with an appealing sweet yet balanced flavour. The plants were stopped at 4 or 5 trusses, and with individual trusses weighing up to 1 ¼lb/567gm each, the result was a good crop of blight resistant tomatoes.

Marmande

Type Semi determinate.   65
Site Outdoor or cool greenhouse.
Fruit Large firm irregular round shaped scarlet tomatoes.
Plus points Heavy early crops, does well in cool weather. F.V.

Marmande, and the improved version, Super Marmande, originally came from France where its resistance to *Fusarium* and *Verticillium* wilts made it a favourite in the Mediterranean area. The plants are semi-determinate and yield large crops of early bright scarlet, rather irregular fruit which are round, slightly flattish, and with ribbed shoulders.

The vines continue to set and swell fruit in cool weather, but when ripened up in hot sunshine the firm meaty flesh of Marmande is delicious.

Roma

Type Compact bush.
Site Cold greenhouse or outdoors.
Fruit Solid flesh suitable for all cooking.
Plus points Good crops of flavourful kitchen tomatoes. V. F. A.

Roma sounds as if it should be Italian, but it was originally bred by the United States Department of Agriculture for the American canning industry in 1955. A few years later the Harris Seed Company of Rochester New York, which was founded by an Englishman, Joseph Harris, in 1879, improved Roma by breeding into it resistance to *Verticillium, Fusarium,* and *Alternaria*.

Roma is one of the best tasting of the

paste tomatoes and, although it can be eaten straight from the plant, the solid thick dry flesh lends itself to paste, soups, sauces and ketchup. The fruit are plum-shaped with a slight neck and are up to 3 x 2 ½"/5x7 cm in size, on compact plants which can yield heavy crops. A single stake and a loop of twine will keep the tomatoes clear of the soil, but a heavier crop results from letting the plants sprawl unchecked over the ground with just a layer of straw beneath them.

Slava

Type Potato leafed bush. 65
Site Sunny outdoor border.
Fruit Sweet bright red.
Plus points Very early, some blight resistance.

Do you seek an early, outdoor dwarf bush tomato with lots of bright scarlet fruit, that will fit neatly into a flower border or any other sunny spot? If so, try Slava.

Originally from Czechoslovakia, where its name 'Slava' means 'Glory', this little beauty will glorify any border in mid-summer with its glorious display of 1-2oz/30-60gm tear-shaped fruits. Slava tomatoes are sweet with enough acid to provide the sort of tang that means when planted alongside a path the fruit never reaches the house.

Sub Arctic Plenty

Type Dwarf bush   50
Site Outdoors.
Fruit Bright orange red, solid and fruity.
Plus points Will grow outdoors in cold districts.

Dwarf outdoor varieties that will thrive in the colder parts of the country are always welcome and Sub Arctic Plenty is one of the best. Of the 4 Sub Arctic varieties of tomatoes, 2 are available in the UK. The other one is Sub Arctic Maxi.

Sub Arctic Plenty, a favourite over

the last 60 years, is hardier than most dwarfs and will set fruit under comparatively cold conditions. The bright orange red tomatoes, about 2"/5cm in size, ripen in the early summer sun with a sweet, acid delicious flavour. Around 9lb/4kg of fruit can be expected from each plant, which can be tied to a stake with twine or left to sprawl over the ground, preferably with a handful of straw underneath to keep the fruit clear of the soil. Being a compact plant, Sub Arctic Plenty can be spaced from 12-18 "/30-45cm apart, yielding a heavy crop from a small area of ground.

Sub Arctic Maxi

Type Dwarf. 55
Site Outdoors.
Fruit Sweet, juicy 2-4 ounce,/56-113gm in weight.
Plus points Early outdoor bush that does well in the North.

Sub Arctic Maxi is the other Sub Arctic variety that does well in northern areas. Slightly later than Sub Arctic Plenty and not as numerous, the fruit of 'Maxi' are larger and once ripe will hold longer on the vine in good condition; this is a very real asset to allotment gardeners who can only garden once or twice a week. The dwarf plants carry clusters of bright red sunburn resistant, firm but juicy tomatoes. Although the vines are trouble free and can be left to grow where they will, the fruit is cleaner when the plant is tied to a stake or grown in a pot.

Tornado

Type Compact dwarf.
Site Outdoors.
Fruit Sweet and juicy with a thin skin.
Plus points RHS Award of Garden Merit. Very early.

Many dwarf tomatoes ripen all their fruit over a matter of days or a week or two at most, which is fine for an early catch crop or in a very short summer district.

Tornado, however, will ripen its fruit regularly over 3-4 months, providing a steady supply of glowing red, thin skinned,

1"/2.5cm sweet juicy tomatoes. This is one those varieties specially bred to make the most of the variable British summer.

The neat compact plants are ideal for pots or tubs which can be moved around into the sunniest spot in the garden as the summer progresses to keep up a constant supply of fresh fruit.

Lycopersicon pimpinellifolium

Type Sprawling multi stemmed bush.
Site Cold greenhouse or outdoors.
Fruit The smallest of all the currant varieties.
Plus points One of the original wild species. Living history. Sweet tasting.

A wild tomato among a list of cultivars? Yes indeed. Lycopersicon Pimpinellifolium – LP – is worth growing for several reasons. As one of the original wild species of tomatoes, it is a piece of living history itself, being the forebear of so many of the small cherry varieties.

For anyone interested in creating their own personal tomato variety LP with its readily pollinated open flowers is easy to use for breeding new varieties. With good disease resistance, less acidity and a high level of vitamins and minerals, descendants of LP make home tomato breeding a fascinating and sometimes rewarding hobby. The tiny fruit of LP are sweet and children find them enticing, making them an excellent introduction to tomato eating.

CHERRY TOMATOES

Gardeners Delight

Type Very vigorous cordon.
Site Heated or cold glasshouse, or outdoors.
Fruit Large trusses of 1"/2.5cm sweet, crack resistant fruit.
Plus points Easy to grow, consistently good crops.

When the Royal Horticultural Society gave Gardeners Delight an Award of Garden Merit it was for a vigorous plant carrying sweet cherry sized fruit. Since that award Gardeners Delight has gone on to become one of the most widely grown of all tomato varieties. For as long as the season's growing conditions continue ' Delight' will set and swell, long, double and treble trusses making it a prolific cropper.

Gardeners Delight was the subject of extensive research at the Glasshouse Crops Research Institute where it was found that the best flavoured fruit came from plants that had been grown in soil rather than peat based composts and kept just on the dry side of moist.

Golden Cherry F1

Type Cordon.
Site Greenhouse or outdoors.
Fruit Thin skinned and very sweet.
Plus points Prolific crops with superb flavour.

Golden Cherry is not just another yellow tomato. With all the power of an F1 hybrid this really vigorous variety produces prolific, long trusses of widely spaced easy to pick fruit. The golden yellow cherry sized tomatoes are thin skinned and packed with a delicious sweet tangy aromatic juice.

Ildi

Type Strong tall cordon.
Site Hot or cold greenhouse.
Fruit Very small, very sweet.
Plus points Huge crops of cocktail sized cherries.

This variety has trusses covered with hundreds of tomato flowers which turn the greenhouse into a flower garden; fortunately they don't all set, for if they did they would pull the trusses off. When it comes to fruit numbers Ildi is in a class by itself, the plants are strong growing and need well supporting. Not only the plants but some of the huge fan shaped trusses may need string supports to help them carry a vast crop of small fruit.

Cocktail or currant tomatoes are only fingernail size ½oz/14gm in weight, but what they lack in size they make up for in flavour, juicy, sweet and with a hint of 'tang'. Children love them and they look simply enticing when scattered over a salad.

Loveheart F1

Type Strong cordon.
Site Hot or cold greenhouse.
Fruit Bright cherry with heart shaped fruit.
Plus points Prolific consistent crops.

Loveheart, or 'Cutie' as it is known, was introduced in 2005, offering heart shaped cherry tomatoes with a true tomato 'tang'. Carrying 6 trusses in the first 5 feet/1.5mtr of stem the vines grew to 16 feet/ 4.8 mtr over a 6 month season, carrying 14 trusses.

The tomatoes are 1-1¼"/2.5-3cm in size, bright red, thin skinned with a vibrant 'tang', and although prolific not all fruit were heart shaped. The hybrid vigour built into Loveheart lends itself to growing as a roof crop. Once the plants reach the top of their supports they are trained along the wires or, where available, over the roof cross ties where the tomatoes hang down like bunches of grapes. Roof crops are very productive but all side shoots need to be removed while they are still small to ensure good air circulation to avoid fungus diseases.

Mini Charm F1

Type Strong vigorous cordon.
Site Hot or cold greenhouse or sunny outdoor border.
Fruit Very sweet full bodied cherry.
Plus points Has good disease resistance to V. F 1&2. TMV.

Over a period of time glasshouse soils growing a range of crops may become contaminated with one or more fungal diseases and when tomatoes have to be repeatedly grown in them the crops deteriorate. Replacing the greenhouse soil is one way of dealing with contaminated soil but for several reasons it is not always practicable, and disease resistant varieties may provide a useful alternative. Mini Charm has inbuilt resistance to V. F 1&2 and TMV and with plenty of hybrid vigour does well on doubtful soils.

Long trusses of bite sized mini plum shaped tomatoes with a lovely sweet flavour make Mini Charm a firm favourite at tomato tastings.

Ocradel

Type Strong cordon.
Site Cold greenhouse or outdoors.
Fruit Sweet acid and juicy.
Plus points The 'Del' series is available in 11 colours.

In the days when the Glasshouse Crops Research Institute was at its most productive, the tomato breeding programme was in the capable hands of Mr Lewis Darby, who then went on to develop the 'Del' series of tomatoes.

Using Gardeners Delight as a source, careful selective breeding resulted in the colourful and delicious 'Del' series. Red, Chocolate, Golden striped, Canary yellow, Deep orange, Dainty blush pink, Red brown with green stripes, Very pale lemon, Dusky pink, Easy peel red and Ocradel, the ochre one as illustrated. All the plants are vigorous, producing a succession of trusses, with fruit less prone to greenback and splitting than Gardeners Delight.

Mixed seed is available and the tall plants can be trained into a most attractive roof crop hanging with different coloured sweetly acid fruit simply asking to be picked and eaten. A salad bowl of mixed 'Del' series tomatoes is a feast for the eyes as well as the taste buds.

Quentin F1

Type Very vigorous single or double cordon.
Site Cold greenhouse or outdoors.
Fruit 1"/2.5cm 'greenback' type round fruits.
Plus points Excellent flavour, early cropper.

Quentin must be one of the best new introductions of 2006. It is sweet with just the right balance of tangy acid and holds its flavour right through the season. In 2005 in my outdoor trial this variety started to ripen in early August and for 3 months provided a continuous supply of ripe fruit of excellent quality and flavour.

Although it performs so well outdoors Quentin was bred in Norfolk specially for growing in home glasshouses where from an early planting its F1 vigour carries it through to late autumn. The well spaced leaves and open habit of the plant allow plenty of air movement contributing to a healthy plant and steady ripening of luscious fruit.

Red Pear

Type Tall lax vine.
Site Hot or cold greenhouse.
Fruit Small true pear shape of about 1½ to 2"/4-5 cm long.
Plus points Historic variety with masses of 'Red Pears'.

Along with Yellow Pear, Red Pear is one of the oldest tomatoes still in cultivation, being mentioned in gardening literature 200 years ago. The plant is a an untidy sprawling vine that needs well supporting to carry a heavy crop.

A 5 foot/1.5 mtr length of vine will carry 5 large fan shaped trusses covered with up to 100 flowers, many of which will set and turn into masses of small 'Red Pears'.

The fruit is solid, often crunchy and of a mild flavour and very much a novelty.

Sometimes called the 'Fig Tomato', the Pennsylvanian Dutch are said to dip Red Pear in a hot sugar and tomato juice syrup and then sun dry them.

Ruby F I

Type Strong vigorous cordon.
Site Greenhouse or sunny outdoor spot.
Fruit Bright red ovals with a luscious sweet flavour.
Plus points Prolific, setting up to 30 fruits per truss.

Look at the prolific growth of a Ruby tomato plant, taste the superb flavour of its fruit and it is easy to see why this F1 hybrid was given the RHS Award of Garden Merit.

The trusses are long and single carrying anything up to 30 ½oz/15gm size striking red oval tomatoes with a succulent texture and a full bodied tantalizing flavour.

Sakura F I

Type Strong growing cordon.
Site Greenhouse, hot or cold.
Fruit Deliciously sweet.
Plus points Prolific cropper on F. and TMV. resistant plants.

Long, fully set, enticing trusses hang from the stem of Sakura tomato plants just asking to be eaten. Even in a cold greenhouse this high yielding variety keeps on steadily growing. With relatively small 10"/25cm long leaves and 8-10"/20 –25cm between trusses Sakura can pack 6 bunches into 5 foot/1.5mtr of stem length and still leave room for warm air to circulate.

The cherry sized brick red tomatoes are full of sweet rich juice delightful to pick and eat when passing.

Ship Saint F1

Type Cordon.
Site Inside or outdoors.
Fruit Very sweet midi plum shaped.
Plus points Tolerant to V. F. C. TMV. heat, rain, and cracking.

This F1 midi plum shaped tomato has been specially bred to produce a good crop where soils are suspect or some disease is known to be actually present. It has shown good resistance to the diseases listed, with the added attractions of tolerating high temperatures and when grown outdoors is not prone to crack or split after heavy rain.

The midi plum shaped tomatoes are particularly sweet and weigh in at around 1oz/28gm each and look extremely attractive on the salad bowl.

Snapper

Type May be grown as a single, double, or triple cordon.
Site Cold glasshouse or outdoors.
Fruit Large olive or small plum shape, crisp with fine flavour.
Plus point Very vigorous plant that just keeps on growing and producing.

Snapper is new for 2006, being the product of a highly successful breeding programme to create a tomato plant that is easy to grow either under cover or outdoors, with an open habit and built-in longevity. A plant that will yield fruit with an excellent flavour, that will travel whole without bruising in a lunch box yet is highly desirable for the dining table.

Snapper tomatoes are oval to small plum in shape, a glowing red in colour and having crisp flesh with an enticing taste. Here in Yorkshire they cropped through

into late autumn and the last ones ripened readily off the vine, and kept well when dried. The plants were slow to start but then turned into vigorous stringy vines. If stressed, however, some plants may stop growing at the 5th node; if this should happen let one or more shoots take over as leaders. A sister to Snapper is Sultana which has similar characteristics but the fruit are perhaps a little smaller, with an intense flavour and if possible an even brighter shade of red.

Sun Baby

Type Strong growing cordon.
Site Hot or cold greenhouse.
Fruit Bright clear yellow, with a juicy well balanced flavour.
Plus points Early to ripen, a consistent cropper.

One of the earliest greenhouse yellows to start ripening, Sun Baby continues to provide top quality 1-1¼ "/2.5 -3.5cm sized tomatoes for over 4 months. The growth is vigorous for such a neat and compact plant which manages to cram 6 , 6 ",/15cm trusses into 5 feet/1.5mtr of stem. The short leaves allow plenty of warm air to circulate keeping the plant healthy and speeding up ripening, while allowing for closer planting than some of the larger leafed varieties. An altogether neat, productive and trouble free variety.

Sun Gold

Type Strong, at times almost rampant.
Site Hot or cold greenhouse or outdoors.
Fruit Very sweet bright orange cherries.
Plus points Prolific variety, very long season of super sweet fruit.

When Sungold was first introduced it was one of the first super sweet small tomatoes to change people's views on yellow fruit. It is consistently sweet over up to a 6-month season. One year, during early December I picked ripe fruit from the 20th truss in a cold greenhouse here in Yorkshire from vines that had not stopped growing. It has to be said that when fully ripe the thin skinned fruit has a tendency to split so needs picking a day or so earlier than when fully ripe.

For gardeners who seek a low acid tomato, a larger fruited offspring of Sungold is Sungella, more orange in colour than its parent it has lower acidity yet is still juicy and sweet.

Sweet Million

Type Very vigorous cordon.
Site Inside or out.
Fruit Super sweet cherry.
Plus points Prolific cropper. Resistant F1. TMV. St.

Her is an easy variety to grow, with excellent setting qualities Sweet Million is almost guaranteed to produce a huge number of 1-1½"/2.5-3.5cm sized fruit.

The very long trusses overlap to form a chain of cherries belonging to the 'pick one eat one' school of tomato harvesting. Early maturing, the plants will crop heavily in greenhouses or yield super sweet outdoor fruit right up until the first frosts.

Sweet Olive F1

Type Strong cordon.
Site Hot or cold greenhouse or outdoors in a sunny spot.
Fruit Olive shaped. Delectable. Striking deep red.
Plus points Easy to grow, open habit, prolific.

Sweet Olive was the most sought-after of all our early greenhouse tomatoes in 2005; one taste and it was 'More please' and the supply just could not meet the demand.

The plant, bursting with hybrid vigour, is tall, the leaves are 10-11"/25-28cm in length and only 7 "/18cm wide and with only a few small side shoots warm air moves freely around the plant and ripening seeds.

Trusses, which may be single, double, or triple, grow every 6-9"/5-23cm up the stem and as they may be a foot/30cm long, they can overlap one another in a solid column of Sweet Olives. The fruit rarely exceed ¾ oz/20gm in weight or 1¼-1½"/3-4cm in length – but what an absolutely superb flavour.

Yellow Butterfly

Type Rampant cordon.
Site Heated or cold greenhouse or sunny outdoor protected corner.
Fruit Hundreds of tiny yellow very sweet tomatoes.
Plus points Sweet and decorative over a long season of cropping.

Yellow Butterfly is one of the strongest growing and most prolific of all the cock-tail sized tomatoes. With only a foot/ 30cm or so between the huge fan shaped trusses, many of them up to 18"/45cm long; these hang down the plant overlap-ping each other creating a dramatic wall of fruit.

If the plants are over watered and fed too much nitrogen, they grow wildly all over the greenhouse and many of the hundreds of bright yellow flowers fail to set. But when Yellow Butterfly is kept moist but not wet and fed with a high potash tomato feed the result is many hundreds of 1"/2.5cm teardrop shaped very sweet delicious fruit.

Whole trusses of late set cocktail cherries can be cut from the plants and hung up in a warm place indoors to provide delicate nibbles right through the late autumn.

STANDARD SIZED TOMATO VARIETIES

Ailsa Craig

Type Moderately vigorous cordon 68.
Site Hot or cold greenhouse or sheltered outdoor area.
Fruit Smooth, uniform scarlet, good flavour.
Plus points Large crops on heavy soil.

Raised by Alan Balch in Ayrshire around 1910 and named after the island off the West Coast of Scotland, Ailsa Craig has headed lists of tomato varieties for over 90 years.

The plant is a lively grower with an extensive root system that thrives in most soils and does very well in heavy soils. The trusses are often long, and may split into 3 or 4 branches around the 3rd or 4th truss level.

The leaves are large and inclined to be soft and fleshy making them predisposed to attacks of leaf mould(*Cladosporium*). But when fed with plenty of potash and moderately watered, Ailsa Craig has stood the test of time as being capable of producing heavy crops with a good flavour. Over the years inevitably different strains develop and it is said that there are at least 30 strains of Ailsa Craig in existence, so go for a strain listed as re-selected, meaning carefully selected seed for the best results.

Alicante

Type Cordon. Indoors 68 days, outdoors 85 days.
Site Cold glasshouse or outdoors.
Fruit Medium sized, round, full flavoured.
Plus points Very easy to grow, early, reliable and greenback free.

For anyone with an outdoor sunny border, or bright corner to place pots or

containers and wanting to grow tomatoes for the first time, Alicante will provide a successful crop. This easy, adaptable variety was given an Award of Garden Merit by the RHS for its consistent reliability in producing attractive medium sized tomatoes which ripen well both under cover and outdoors.

Aromata F1

Type Greenhouse cordon.
Site Greenhouse only.
Fruit Round and solid.
Plus points Does very well in northern Britain. TMV.

With so many tomato varieties bred to take advantage of warmer climes, finding a cultivar specially bred for the European weather is a rarity.

Aromata was specially developed for North West Europe; it is a non-greenback tomato, mosaic resistant hybrid which certainly does well in northern England. Being produced right until the end of autumn, the round solid juicy fruits with a distinctive flavour are consistently on every truss right up the plant.

Arctic

Type Cordon.
Site A sunny outdoor border.
Fruit Medium sized, smooth, with a good flavour.
Plus points An early, consistently good cropper.

Early, outdoor tomato varieties capable of producing heavy crops thrive in a warm sunny spot. Plant Arctic where it will get the sun and it will be loaded with delicious fruit, from medium to small in size ripening to a bright red on a trouble free plant. The smooth skinned fruit ripen up in a steady succession to provide a continuous supply right up to the first frosts.

Aviro F1

Type Cordon.
Site Cold greenhouse or outdoors.
Fruit Medium to large pointed plum shaped.
Plus points Heavy cropping TMV resistant plants.

Aviro used to be known as 'Orange Plum', as it is a delightful orange colour. Grown as a cordon, the deliciously sweet fruit are medium in size, but when grown outdoors and stopped at 4 or 5 trusses they may reach 3"/7.5cm in length and 2"/5cm across. Individual trusses may tip the scales at over 2 lb/1kg in weight.

Orange tomatoes are rich in beta-carotene and Aviro has almost double the amount of Pro-vitamin A and vitamin C as red tomatoes. Very attractive when sliced in salads, they are superb when roasted.

Carters Fruit

Type Strong cordon.
Site Indoors or outside.
Fruit Distinctive: ripens to crimson with a powdery 'bloom'.
Plus points An old variety still worth growing.

When Messrs Carters of High Holborn, London, introduced Carters Fruit over 70 years ago, it was something different; in appearance it reminded gardeners of the old 'Peach' tomato. The skin of this new introduction had a fine texture and was covered with a powdery 'bloom' and when fully ripe was a rich deep crimson colour.

Over the years some strains seem to have lost their characteristic bloom but the sweet luscious flavour remains; the flesh is solid but free from core and with not a lot of seeds Carters Fruit makes a lovely historic addition to the salad bowl.

Cavendish

Type Vigorous cordon.
Site Hot or cold greenhouse.
Fruit Medium sized crimson globes.
Plus points Prolific heavy cropping main crop.

Cavendish was a firm favourite with home gardeners around 70 years ago, because at a time when many cultivars had loose, widely spaced irregular sized tomatoes on their trusses, Cavendish was 'short jointed'. The compact trusses carrying uniform fruits grow every 6"-8"/15-20cm right up the stem.

Mainly grown now by enthusiasts as a link with the past, the plant produces a continuous supply of high quality fruit of a reddish-crimson colour, with a thin skin and a juicy old fashioned tomato taste.

Costoluto Fiorentino

Type Cordon 80
Site Hot or cold glasshouse or outdoors.
Fruit Vivid bright red, slightly ribbed, variable oval- round.
Plus points Flavour; grows well over most of the country.

Costoluto Fiorentino is an heirloom variety from the Tuscany region of Italy, with similar characteristics to Costoluto Genovese. It is a tall plant with long leaves and some 6"-7"/15-18cm between trusses.

A typical plant will crop over a long season and carries up to 4 trusses over any 3 foot/1 mtr of stem. The stubby trusses carry up to 4 or 5 slightly ribbed vivid bright red fruit. In Yorkshire the first cold house Costoluto started to ripen in late June, and continued to crop steadily until mid-October; the fruit are juicy with a high sugar and acid content creating a succulent delectable taste either fresh or cooked; they are exquisite when picked fully ripe and roasted.

There is an outdoor sprawling version of Costoluto Fiorentino which is a good cropper and while it does not need supporting a layer of straw under the trusses helps to keep the lovely fruit clean and free from soil splashes.

Dombito F1

Type Cordon. 67
Site Hot or cold glasshouse or outdoors.
Fruit Large, round or slightly flattened oval.
Plus points Early heavy crop, with a good rich flavour.

This hybrid was originally bred for the commercial grower to market as an improved beefsteak type yielding high quality fruit. Dombito tomatoes weigh up to 10oz/283gm and are up to 3½"/9cm in size. Some of the fruit are slightly oval in shape and are very solid, meaty and have a satisfying full bodied flavour.

Where space is limited, Dombito will fill it with a compact, short jointed plant growing 5 trusses in 5 feet/1.5 mtrs. A full set on each truss would pull the plant down, so expect 3 to 5 solid fruit per truss which can still add up to a heavy crop.

Druzba

Type Cordon, late.   80
Site Indoors or when planted outdoors, stop early to harvest a crop.
Fruit Large smooth, sweet and juicy.
Plus points Heavy crops with disease resistant foliage.

With its name translated from the original Bulgarian, the heirloom variety 'Druzba' means ' friendship' and it is easy to see why a gift of these glorious globes is always welcome among friends.

The plant is capable of producing

heavy crops of smooth, round, bright scarlet fruit which may reach anywhere between ½–1lb/250–500gm in weight. Combining a fine mixture of rich sweet juicy flesh with a slightly tart tang the tomatoes look most attractive set against their disease resistant foliage.

Essex Wonder

Type Very strong cordon.
Site Indoors or out.
Fruit Round and firm.
Plus points Vigorous on most soils.

When Messrs Dobie and Company introduced Essex Wonder, it quickly became known as the best variety for growing in cold houses and out of doors. Its very popularity encouraged many trade growers to save seed and not everyone was careful enough to keep up a high standard of selection, with the result that several strains developed in different parts of the country.

The original Essex Wonder is an extremely vigorous plant, in a cold greenhouse it can quickly become rampant when planted in loose soils, and needs a firm soil and restricted root run to keep it under control. Outdoors when planted in firm soil its many stubby trusses carry round, orange-red solid fruit with a good balanced flavour.

Ferline F1

Type Medium vigorous cordon.
Site Cold greenhouse or outdoors.
Fruit Deep crimson with a jewel like glow.
Plus points Has very good resistance to V. F2. M. S.

Over the 10 years that I have been growing this French hybrid it has proved a consistently heavy cropper particularly outdoors. The weather pattern is changing and attacks of late or potato blight, *Phytophora infestans,* are occurring regularly across the country causing widespread crop losses. Ferline has very good resistance not only to blight but to *Verticillium* and *Fusarium* soil borne wilts.

Outdoors its large healthy root system quickly supports a strong plant carrying 6 or 7 trusses before being stopped; in most seasons at least 5 of these trusses will ripen up on the plant with the mature green fruit ripening up indoors to the very last one. The tomatoes are solid and meaty with a lovely balance of sugar and acid and a distinctive tang and will keep firm for some time after being picked, - if allowed to.

Global

Type Vigorous cordon.
Site Hot or cold greenhouse.
Fruit Sweet juicy globes.
Plus points Prolific cropper with a good flavour.

The name is a play on its shape. Global is a lovely bright orangey red globe.

Thin skinned and full of juice, with a good balance of sugar and acid, the often double trusses of 10 to 12 fruit ranging from small to medium grow on a compact plant of medium vigour.

Golden Sunrise

Type Moderately vigorous cordon.
Site Hot or cold greenhouse.
Fruit Smooth bright yellow.
Plus points Consistent cropper with less acid fruit.

This variety introduced by Messrs. Carters many decades ago was awarded a First Class Certificate by the RHS and quickly became a popular favourite. The bright golden yellow 2"/5cm tomatoes taste less acid than many red cultivars, and with attractive solid flesh have a sweet, mild, tempting flavour which many people consider to be still the best

yellow variety.

Trusses carrying 6 to 8 medium sized fruit can grow as closely together as 7 on a 5 foot/1.5mtr length of stem and will continue to set well until the plant is stopped. Although it will grow outdoors, with a thin smooth skin, Golden Sunrise produces better quality fruit when grown under cover.

Gourmet

Type Glasshouse cordon.
Site Hot or cold greenhouse.
Fruit Medium sized scarlet with a good flavour.
Plus points Vigorous variety, consistent, continuous cropper.

Trouble free, early fruiting tomato varieties that just keep on growing are the mainstay of a long season's crop. Gourmet, with an ability to carry up to 15 trusses over the season, is one of many suitable varieties available when an F1 hybrid with disease resistance is required.

Gourmet has an open habit which allows for good air circulation among the plants and ripening fruit. The tomatoes, which are carried on 6"/15cm long trusses, are bright red to scarlet in colour, medium in size, firm but juicy with a good acid to sugar balance ensuring a tempting full bodied flavour.

Marglobe

Type Strong cordon.
Site Hot or cold greenhouse, outdoors in a sunny spot.
Fruit Solid, round and sweet.
Plus points Resistant to *Fusarium*. Still available after 90 years.

The pedigree of Marglobe goes like this: The variety Marvel(1918) was descended from the old variety Stone (1889). Globe (1905) was the offspring of the legendary Ponderosa(1891)

When the United States Department of Agriculture were seeking a *Fusarium* resistant variety back in 1925 they crossed Marvel and Globe and the result was Marglobe. This variety quickly became a favourite with home gardeners because of its appearance and flavour.

The tomatoes ripen to a smooth scarlet red with an appealing flavour tending more towards the sweet than the acid. Being thick walled the fruit are sometimes subject to radial cracking, a characteristic that was bred out of the original when Messrs Ferry-Morse developed an improved version named Marglobe Supreme.

Marion

Type Cordon.   70
Site Hot or cold greenhouse.
Fruit Medium large glowing red.
Plus points Disease resistant. Stays firm for 2 or more weeks after picking.

Tomato aficionados try to keep a plant fresh supply of their favourite fruit ripening as far into the autumn as conditions allow. Equally welcome are trusses

of mature green tomatoes to be ripened indoors to grace the Christmas table and hopefully well into the New Year. This calls for strong vigorous varieties, able to grow against the falling daylight levels and capable of withstanding autumnal fungus attacks, while still yielding good quality fruit. Marion neatly fills this spot.

Grown as a main or late crop, the bright orange red fruit light up the autumn and when picked will remain firm for 2 or more weeks, lovely and ripe. At the end of the season, Marion's 2"/5cm tomatoes will store well and can be ripened a few at a time in a warm room. We have eaten Marion as late as early February thus ending a season that began with Stupice and Sioux in mid-April, almost 10 months of home-grown tomato eating.

Moneymaker

Type Tall strong cordon.   70
Site Hot or cold glasshouses. Outdoors.
Fruit Medium to large, scarlet red.
Plus points Easy to grow, heavy cropper. Greenback free.

When Mr F. Stoner of Southampton began breeding tomatoes a century ago, many of the cultivars then being grown had 'greenback' (the condition where the ripening fruits had green, often hard shoulders), so his aim was to breed greenback free, superior quality round fruit.

Southampton bred Gradewell and Stoners MP were good while Exhibition was probably the showman's favourite, winning countless shows up and down the land.

But it is Stoners 'Moneymaker' that has really stood the test of time and over the years has probably been the most widely grown tomato in home gardens.

Moneymaker is tall, strong and easy to grow; setting well in spring, it produces heavy crops of smooth round tomatoes of a brilliant scarlet red. If the plants are grown 'soft' with too much available nitrogen, the large fleshy leaves are prone to leaf mould and the fruit, although looking good, loses its flavour. When the plants are fed with a high potash feed however that old fashioned tomato taste is back and with harder leaves the foliage stays healthier longer.

Monte Carlo

Type Cordon.
Site Outdoors,
Fruit Round greenback. Intense red.
Plus points Prolific and disease resistant.

A disease resistant cordon with medium sized round fruit is rather unusual among modern Italian tomatoes but Monte Carlo is one of them. This intense scarlet variety with a fine 'Italian' flavour does well outdoors in our climate; overlapping trusses carry 6 to 8 medium sized fruit containing juice with a delectable mixture of sugar and acid.

Olivade F1

Type Medium vigorous cordon.
Site Inside or out.
Fruit Oval, plum shaped, excellent flavour.
Plus points Good resistance to V. F1&2 C3 TMV.

Olivade is a good choice when an early ripening juicy, fleshy plum-shaped tomato is needed for both eating fresh and cooking. When fully ripe the 3½ oz/100gm fruit develop a luscious full bodied flavour good enough to warrant the RHS Award of Garden Merit.

The plant has medium vigour and such a selection of disease resistance that it can be grown almost anywhere, cropping very well in cold greenhouses and carrying at least 4 to 5 good trusses yielding 6 to 8 fruits each when grown outdoors.

Orange Banana

Type Cordon   85
Site Hot or cold greenhouse or outdoors.
Fruit Beautiful orange colour, sweet and solid.
Plus points A most attractive paste and soup type fruit.

Orange Banana has an immediate appeal as it hangs in clusters of beautiful fruit. This is one of the paste type of tomato and while making a lovely addition to the salad bowl they are ideal for cooking and excellent for drying. The tomatoes are 3-4"/8-10cm long with pointed ends, they are sweet and have a slight citrus undertone.

Pantano

Type Cordon.
Site Hot or cold greenhouse. Sunny outdoor border.
Fruit Large slightly scalloped and tasty.
Plus points Prolific, prolonged ripening season, beefsteak.

Pantano, also known as Pantano Romanesco, is a very old Italian heirloom variety.

A rather untidy looking vine, it carries its fruit well spaced out along the truss allowing plenty of air to circulate. The tomatoes ripen steadily along the truss providing a constant supply of solid fleshy fruit, slightly scalloped or rippled with few seeds and a tasty flavour.

Plumpton King

Type Vigorous cordon.
Site Inside or outdoors.
Fruit Smooth round tasty globes ripening to a dark red.
Plus points A personal favourite 55 years ago, still being grown today.

During the 1930s and 40s tomato breeders were keen to develop varieties capable of producing heavy crops of good quality fruit that appealed to both commercial and amateur growers. The variety Market King was just such a one and quickly became popular, particularly in the North of England. Then Market King was improved to yield even better crops and Plumpton King was born. It was some 55 years ago that I first grew Plumpton King for the quality and quantity of the medium sized deep red fruit with the true tomato taste, but over the years amid a host of new varieties our strain of Plumpton King was lost.

A decade ago Mrs R. Sutcliffe, the

widow of Mr Harold Sutcliffe of Bingley, gave me his seed collection. For 36 years Harold had diligently saved seed from the best fruit of each years crop of Plumpton King keeping the strain with all its original characteristics. I gave a good proportion of the seed to the Henry Doubleday Research Association Heritage Seed Library who offer them to members from time to time, so Plumpton King is once again being grown and enjoyed.

Principe Borghese

Type Cordon.
Site Cold greenhouse or outdoors.
Fruit Plum shaped red, carried in clusters.
Plus points A cook's tomato, dries well.

An old heirloom variety from Italy. Principe Borghese is the gambler's tomato. A wonderful variety for splitting in half and drying, but it is a very real gamble whether or not there will be enough hot sun to dry it out naturally when cropping is at its peak.

The tomatoes are the shape and size of an egg and with little juice or seeds dry out to a savoury candy. Is it worth the gamble? of course it is, there is bound to be some hot sunshine at some part of the season, and even if the weather does not co-operate Principe Borghese makes a magnificent roast tomato soup.

Rose de Berne

Type Vigorous cordon.
Site Inside or outdoors.
Fruit Large dark pink, tastes similar to Brandywine.
Plus points Connoisseurs' choice for flavour.

Rose de Berne, originally from France, is one for the connoisseur. The mainly round smooth tomatoes of most strains have a very thin skin and the soft tender flesh is juicy and delectable. Occasionally tomatoes as shaped in the photograph may appear. The fruits are around 6-8oz/170-230gm in weight and when the bottom trusses are carrying a heavy load the next 2 trusses may be lighter, but then the plant resumes steady cropping.

Rose de Berne grows well in most soils especially in a cold greenhouse, where the thin skinned beauty needs careful picking and handling.

Rudolph F1

Type Tall strong cordon.
Site Cold greenhouse or sunny spot outdoors.
Fruit Small, bright red, plum shaped.
Plus points Vigorous plant with rich fruity tomatoes.

From out of the East came Santa and from the same breeders came Rudolph.

As would be expected from an F1, Rudolph is tall strong and prolific, with leaves up to 18"/46cm long and 12"/30cm wide and a truss every 12"/30 cm up the stem.

The long, sometimes double, trusses, carrying up to 25 tomatoes, overlap each other creating a dramatic wall of fruit. Each tomato is about 1½ inches/4 cm in size, a glowing red, small plum shaped delectable blend of sugar and acid.

San Marzano

Type Cordon.
Site Cold greenhouse or outdoors.
Fruit Cardinal red, pear-plum shaped
Plus points Makes superb paste and purée.

This old Italian heirloom is also known as 'Italian Canner' and is very much a cook's tomato. Although Roma super-seded it, San Marzano while later to ripen has drier fruits which reduce to a glorious tomato paste, passata or purée.

A new bush version of the variety, San Marzano Astra F1 with resistance to *Verticillium*, *Fusarium* and fruit rots is now available; it still retains that lovely flavour.

Scotland Yellow

Type Strong cordon.
Site Cold greenhouse or outdoors.
Fruit Sweet and juicy.
Plus points Medium early, continuous cropper.

As its name implies, Scotland Yellow does well in the North where the strong plants just keep on growing right until the end of the season. The trusses are short with 4-6 tomatoes held in a bunch close to the stem which is an asset when planted outdoors in the North where ripening starts around mid-summer and continues right up to the first frosts. Scotland Yellow is a sweet, juicy, firm tomato appreciated for its subtle yet definite flavour.

Shirley F1

Type Strong cordon.
Site Hot or cold glasshouse or sunny spot outdoors.
Fruit High quality, smooth and round.
Plus points Easy to grow heavy crops. Resistant to V. F2. C5. TMV.

With a well deserved Award of Garden Merit and a National Institute of Agricultural Botany recommendation, Shirley is one of the most popular of all varieties, probably producing more quality tomatoes for the home gardener than any other cultivar.

It is an early, short jointed type and to see a good crop of Shirley is to see a solid wall of fruit that continues to thrive even in periods of poor weather. The F1 hybrid vigour combined with the inbred disease resistance allows Shirley to succeed on soils where

other varieties struggle. As with many a heavy cropper, regular high potash feeds are needed to maintain the flavour which can become rather weak when too much plain water is given.

Sioux

Type Compact cordon.
Site Hot or cold greenhouse.
Fruit Small to medium, exquisite flavour.
Plus points Very early: 16 weeks from seed sowing to first fruit.

Over the last 15 years Sioux has become a family favourite, providing those, 'oh so welcome' first ripe tomatoes at the end of April and early May with such a sublime taste.

Originally bred for the great plains of the USA, Sioux has consistently grown well here in Yorkshire where the compact habits of the plant are better grown under glass. Each truss carries some 5 to 8 perfectly round small to medium sized tomatoes with a sublime mixture of sugar and acid.

Although it is a very consistent cropper right up the plant, the total weight of fruit is not as heavy as many glasshouse varieties, but what it lacks in crop weight is more than compensated for by the continuous production of those delicious early tomatoes.

Sonato

Type Vigorous cordon
Site Hot or cold glasshouse
Fruit Orange red globes
Plus points Consistent long season's cropper keeps well after picking

Sonato with its open habit may appear to be rather lax but the vine keeps on steadily producing trusses of 8-10 tomatoes. The orange red fruit is of a consistently high quality, with meaty yet juicy well flavoured flesh. This is one of those varieties originally bred for the market grower but does equally well in the home greenhouse; it performs well with a good crop whatever the season.

Stupice

Type Potato leafed cordon.   Grows to 4–5 feet/1–1.5mtrs
Site Hot or cold greenhouse. Also grows well on a sunny spot outdoors.
Fruit Mid red. Excellent flavour tangy and juicy.
Plus points Very early, with some resistance to potato blight.

Stupice is one of the most constant of the early varieties of tomatoes that I grow.

When sown in early January the first flowers appear 7–8 weeks later and even under poor light conditions their viable pollen ensures a good set. Starting at the end of April the fruit ripen to a mid red with orange undertones, later in the season there may be a tendancy for green shoulder to appear on some of the larger fruit. The ripe tomatoes have a deep rounded flavour with a good balance of sweetness and acidity, weighing between 1–2 oz/28–56gm.

Czechoslovakia is the original home of Stupice, where its resistance to cold combined with excellent flavour, high yields and some blight make it a local favourite. This potato leafed variety reaches a height of some 4–5 feet/1–1.5mtr before it naturally stops itself. The stem is slender and needs to be frequently tied to a cane or grown up strings to support the 7–8 lb/3 – 3.5 kg of fruit the plant is capable of yielding

St Pierre

Type Cordon
Site Cold glasshouse or outside.
Fruit Medium large, bright red, well flavoured.
Plus points Introduced in 1880 and still going strong.

With a name like St Pierre this old heirloom variety couldn't be anything else but French. Developed 126 years ago, this mid to late season variety just keeps on cropping until the frosts. The bright red tomatoes are firm but juicy with a good

balanced flavour, and keep well off the plant either picked ripe or when picked mature green and stored to extend the season by ripening up indoors.

Striped Stuffer

Type Cordon.
Site Cold greenhouse or outdoors.
Fruit A most attractive bi-colour.
Plus points Long keeping kitchen tomato.

Among the multi-coloured types of tomato, Striped Stuffer is an attention seeker.

The medium to large fruits are best described as 'blocky' tending towards the sweet bell pepper in shape and with most attractive red and yellow stripes.

Although it undoubtedly adds colour to the salad bowl the fruit is better stuffed with a mixture of choice and baked in a hot oven. An added bonus in the kitchen is the shelf life, up to 4 weeks after picked ripe and when picked mature unripe they will ripen up slowly over the next 2 months.

Tigerella

Type Cordon 55
Site Hot or cold glasshouse or outdoors.
Fruit Attractive 'Tiger' striped red and yellow.
Plus points Early and prolific with good disease resistance.

With an RHS Award of Garden Merit to its name, Tigerella, which is also known as Mr. Stripey, is the earliest and brightest red of the 'tiger' striped types of tomato.

The stem is thin and wiry with light open foliage interspaced with a truss every 7" to 9"/18-23cm; the plant may appear to lack vigour, but it grows and crops consistently well until stopped. Tigerella tomatoes are a vibrant combination of red and yellow stripes, and although smaller, at 1¼"-2"/3-5cm in size, they pack a distinctive tangy taste inside a colourful skin.

Vanessa F1

Type Strong cordon.
Site Inside or outdoors in a sunny spot.
Fruit Long keeping, juicy and sweet, bright red round fruit.
Plus points Long keeper, tolerant of drought.

Vanessa, given an Award of Garden Merit by the RHS, is one of the new vine-ripening types of tomato. The whole truss ripens up within a few days and when cut from the plant the fruit stays firm for up to 2 weeks after turning red. A useful variety for taking home grown juicy sweet tomatoes away to eat when on holiday.

The plant has good drought and irregular watering tolerance, which is another useful attribute if plants have to be left unattended at holiday time.

Victory

Type Cordon   72
Site Cold house or outdoors.
Fruit Slightly flattened, unique flavour.
Plus points Delicious Russian heirloom from the Sakaline Island.

There have been 2 English varieties named Victory but this Russian heirloom, although sharing the same name, offers a very different fruit.

The 4-6 oz/113-170gm tomatoes are slightly flattened and are an orangey red when ripe, with a unique flavour which is sweet yet tangy, with a curious lemony undertone. The trusses hold their fruit in a bunch and uneven watering may produce slight cracking but this does not in any way impair their special flavour.

Yellow Perfection

Type Potato leaved cordon.
Site Inside or outdoor.
Fruit Bright yellow, mild flavour.
Plus points A slice of history, 1890 vintage.

One of the oldest varieties still in culti-vation, Yellow Perfection is noted for its bright yellow fruits of the highest quality. The prolific tomatoes are small but are among the earliest to ripen and with a mild, delicate but distinctive flavour are appreciated by those with a delicate palate.

BEEFSTEAK TYPE TOMATOES

Beefmaster

Type Cordon.
Site Greenhouse or outdoor.
Fruit Very large beefsteak.
Plus points Good disease resistance to V. F. N. A. St.

This is a variety that needs to be well supported in advance of what is usually a very heavy crop. Individual tomatoes may weigh anything up to 2lb/900gm of solid meaty flesh with a surprisingly good flavour for such a large fruit. This improved hybrid produces heavy yields with a greater tolerance to splitting and cracking, a one slice sandwich filler and a cook's delight.

Big Boy

Type Tall vigorous cordon   78
Site Cold greenhouse or outdoors
Fruit Very large, smooth, round, and crack free.
Plus points Has that true tomato taste.

When W. Atlass Burpee, the American Seed Co., introduced Big Boy in 1949; it was one of those 'new-fangled hybrids' which quickly became famed for its large fruit, which average about 12oz/340 gm each and may reach up to 1lb/454gm in weight.

The variety was improved and gave rise to Better Boy, Early Girl, Wonder Boy and Ultra Girl, all of which have greater disease resistance bred into them. Tomato aficionados however still consider that the bright red, meaty, full bodied fruit of

Big Boy has that true tomato taste.

Compared to present day hybrids, Big Boy may have lower yields, but if two stems are allowed to grow more but smaller fruit are produced. It is a mid season to late variety and to get the best from an outdoor crop needs sowing early enough to plant out a large plant after the last frost has gone.

Brandywine

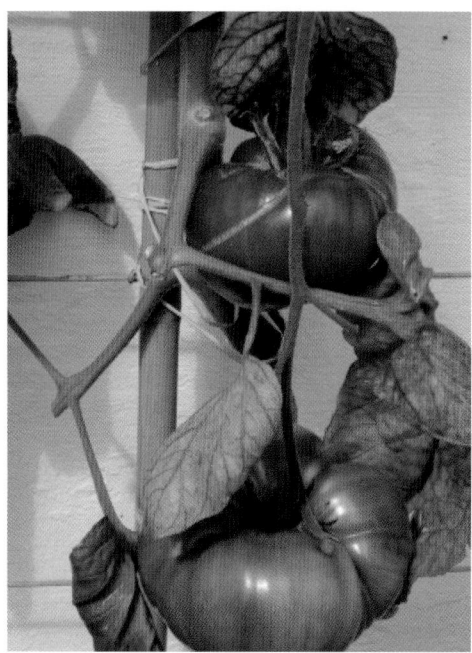

Type Potato leafed cordon   80 or more days.
Site Greenhouse or outdoors in a very sunny spot.
Fruit Very large, from 1-2lb/500g-1kg, delicious.
Plus points A delectable slice of history.

Probably more words in praise of Brandywine have been written in recent years than for any other tomato variety. This large, luscious, pink fruit lends itself to the myths and legends surrounding it. Having seen the magnificent tomatoes grown by the Mennonite community in Canada it is easy to accept that this is an Amish heirloom with deep Germanic roots.

However, Brandywine was introduced by Messrs Johnson & Stokes, a seed firm in Philadelphia, in January 1889 and became popular across the USA. This heirloom variety was a favourite of Ben Quisenbury who maintained hundreds of tomatoes from 1910 to 1960 and bequeathed his collection to Seed Savers Exchange. In 1980 Ben Quisenbury was given a strain of Brandywine by Doris Suddeth Hill whose Tennessee family had grown it for about 100 years and it is said to be the true strain of Brandywine.

At our West Yorkshire Organic Group's annual show, now in its 17th year, everything is judged by taste. When judging Brandywine I have found that the flavour of the exhibits clearly reflect the weather: in a cold wet season the taste is only moderate, but when the fruit matures and ripens in hot sunshine the flavour is truly superb. Care is needed to be consistent when watering the plants as the large fruits are prone to cracking

Burpee Delicious

Type Cordon   78
Site Inside or outdoors.
Fruit Ribbed but smooth and solid. Truly delicious.
Plus points World heavyweight champion.

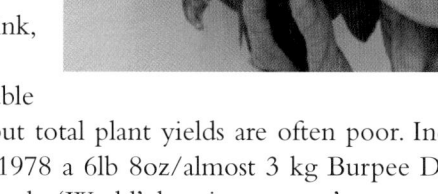

Burpee Delicious is a fine example of the improved beefsteak tomato, a variety that because of its size was destined to carry the name of its breeder, one of America's largest seed companies, W. Atlass Burpee, right round the world

In the early 1950s Burpee began the selection of beefsteak tomatoes leading to the introduction of Burpee Delicious in 1964. Burpee Delicious is a cross between the Yellow Jubilee and a large fruited pink, yet it ripens up to a fire engine red.

The handsome fruit has a delectable flavour with a solid meaty texture, but total plant yields are often poor. Individual fruits may just keep on swelling; in 1978 a 6lb 8oz/almost 3 kg Burpee Delicious entered the *Guinness Book of Records* as the 'World's heaviest tomato'.

Crnkovic Yugoslavian Pink

Type Strong cordon   80 to 90 days
Site Cold greenhouse or sunny outdoor spot.
Fruit Rich juicy pink beefsteak.
Plus points Prolific, disease resistant, non cracking.

This heirloom variety from Yasha Crnkovic of the Vojvod area of Yugoslavia with its large pink-red fruit is as good to eat as it looks. Individual tomatoes can reach 1 lb/500gm in weight with meaty juicy flesh of a rich appealing flavour.

The vines are strong and disease resistant producing a prolific crop of smooth skinned, slightly ribbed, crack resistant fruit.

Cuor di Bue

Type Lax cordon.
Site Hot or cold greenhouse.
Fruit Huge ox hearts, solid with few seeds.
Plus points Reliable moderate crops of large well flavoured fruit.

When I first planted Cuor di Bue some 12 years ago I was not prepared for the sheer size and weight of the fruit which can exceed 1lb/454gm of solid flesh. The plant is of moderate vigour and sometimes has a tendency to shed any unwanted flowers and retard small fruit when it is under the stress of swelling and ripening heavy bottom trusses.

Once the bottom trusses are picked, the plant goes on to set and ripen a few fruit at a time regularly for the rest of the season. The almost seedless tomatoes with a full bodied flavour can be enjoyed fresh, or, the solid flesh is superb roasted, or providing the basis for several mouth-watering dishes when cooked.

Dr Wiche's Yellow

Type Vigorous cordon   80 – 90
Site Cold greenhouse or very sunny spot outdoors.
Fruit Golden yellow, richly flavoured.
Plus points Delicious, productive, American heirloom beefsteak.

How can a Dr Wiche's yellow tomato plant produce such large fruits from so little foliage is the obvious question when looking at this lovely variety. The answer lies in the vigour of the plant; the foliage may be sparse, but the plants are vigorous, long standing and produce heavy crops.

With a slightly flattened shape and a glowing golden yellow colour, Dr. Wiche's tomatoes are meaty and contain few seeds among the delicious richly flavoured flesh.

Giant Syrian

Type Cordon   75 – 80
Site Inside or out.
Fruit Large well flavoured beefsteak.
Plus points Good crops of big fruit.

Trouble free beefsteaks that give steady yields of 1 lb/500gm tomatoes with a good flavour are always welcome. This heirloom variety from West Virginia produces deep pinkish red globe shaped fruits with a smooth slightly ribbed skin. The flesh is solid with very few seeds and a deep rounded flavour.

Hill Billy Potato Leaf

Type Strong cordon   85
Site Cold greenhouse or outdoors.
Fruit Large flattish juicy and sweet.
Plus points A gorgeous, gourmet's fruit.

Hill Billy Potato Leaf is an heirloom variety that hails from Ohio, but its ancestor, West Virginia Hillbilly, originally came from the hills of West Virginia where it reached up to 2lb/1kg in weight.

It is easy to see why this magnificent fruit is sometimes included in the gourmet selection of tomatoes, because of its very sweet tantalising fruity flavour which is low in acid. Often it is late in the season before the huge, slightly flattened heavily ribbed fruits turn to a mottled orangey yellow with red streaks, but they are well worth waiting for.

Kellogg's Breakfast

Type Strong cordon   79
Site Cold greenhouse or outdoors.
Fruit Beefsteak, slightly oblong with smooth ribbed skin.
Plus points Eat fresh or makes delicious juice.

No connection with breakfast cereal at all. This variety is named after Darrell Kellogg of Redford, Michigan, who preserved the variety, having got the seed from a friend. The plants are big, carrying trusses of 4 or 5 fruit which ripen to a bright orange hue. Irregular in shape, some rather oblong with square shoulders and ribbed sides, the tomatoes have a full bodied flavour and make a delicious, bright orange coloured, thick tomato juice ideal for breakfast.

Moskvich

Type Compact cordon   60
Site Outdoors.
Fruit Round or irregular, very delicious.
Plus points Does well in cold areas and cool summers.

Eastern Siberia does not readily spring to mind when thinking about tomatoes, yet that is the original home of Moskvitch. As may be expected, coming from that part of the world, this variety is very tolerant of cool growing conditions and continues to thrive when its warm bloodied cousins are chilled to a standstill during a cold summer.
     Moskovich tomatoes are in the 4–6oz/120-160gm range and are a deep glowing scarlet with a vibrant full bodied taste. Occasionally Moskovich may produce eccentric squarish fruit; with the same lovely flavour they are ideal for neat sandwiches.

Pineapple

Type Strong potato leaved cordon. 85-90
Site Cold greenhouse or very sunny spot outdoors.
Fruit Large delicious bi-colour Kentucky heirloom.
Plus points Beautifully coloured, sweet flesh.

Pineapple is a real beauty. Put part of a slice on open sandwiches and watch them disappear. Only part of a slice because they can reach 2 lb/1kg in weight but even when half that size they are still a huge fruit.

The bright flesh is marbled and streaked through with red, orange and yellow in a most attractive way; juicy and sweet, it is fruity and enticing. Given good growing conditions the strong potato leafed vine will take around 3 months from fertilised flower to ripe fruit, but they are well worth waiting for.

Ponderosa Golden

Type Strong cordon   78
Site Cold glasshouse or outdoors.
Fruit Golden yellow up to 1lb/ 500gm in weight.
Plus points Descended from one of the oldest of all tomatoes.

Ponderosa is one of the oldest names among tomato varieties, going back to 1891 when the Peter Henderson Seed Company introduced the luscious tasting Hendersons 400. The name change came after a 'name that tomato' competition when Ponderosa was selected as the winner.

The original Ponderosa was pink but Ponderosa Red and Ponderosa Golden were developed before 1940. Not as heavy a cropper as some of the newer hybrids, Ponderosa Golden sometimes does surprisingly well. The foliage is rather sparse and in a wet year subject to disease but the large golden yellow smooth skinned fruit are sweet and solid and excellent as one slice sandwich fillers.

Purple and Black Varieties

Black Crimea

Type Cordon   70 – 80
Site Inside or outdoors.
Fruit Round, dark, sweet and attractive.
Plus points Resistant to drought and very productive.

Along with Black Krim and Russian Black' Black Crimea comes from the Black Sea area of southern Russia. This is one of the sweetest 'Black' varieties with the added attraction of being very juicy and slightly sub acid which makes it appealing to children. After a long sunny spell the fruit develops a slightly spicy smoky flavour which disappears later in the season.

When ripe the skin of Black Crimea goes a dark mahogany colour with black shoulders; when cut across the flesh is dark red-brown and is a wonderful addition to the salad bowl.

Black Plum

Type Single or multiple stem semi-determinate.   82
Site Cold greenhouse or outdoor sunny border.
Fruit 1½"-2½"/4-6 cm Black Plum tomatoes.
Plus points Very heavy cropper.

A Russian heirloom variety, Black Plum will stop growing when about 3' 6"/1mtr tall. Grown as a single cordon, the fruit are larger than when grown as a semi-determinate 3 or 4 stem plant. The stems become covered with clusters of plum shaped tomatoes which ripen from a deep mahogany to brown with black shoulders. With crisp flavoursome flesh, they are a one or two bite snack when passing.

Black from Tula

Type Cordon.
Site Inside or outdoors.
Fruit Purple-red – mahogany .Delectable and rich.
Plus points One of best 'Black' varieties, sets well in hot weather.

Black tomatoes may be an acquired taste but one bite of this Russian heirloom and the taste is acquired. A beautiful fruit weighing in at ½-1lb/250-500gm, it blends purple/mahogany/red into a glorious mixture of colours and when cut reveals reddish brown flesh with dark green gel surrounding the seeds. A great favourite at tastings, with its creamy texture and sweet, slightly salty taste, it is quite different to red beefsteaks.

The plants rarely reach more than 5 feet/1.5mtrs in height and are only moderately productive, but the flowers set well even in hot weather, so a steady supply of this magnificent fruit continues to ripen up one at a time.

Cherokee Purple

Type Cordon   80 – 85
Site Cold glasshouse or outdoors.
Fruit Unique dark dusky pink-purple. Luscious flavour.
Plus points Resistant to leaf spot it is one of the best large fruit.

With a name like Cherokee there just has to be a North American connection with this 1890 heirloom. The variety was introduced through the Seed Savers Exchange in 1991 by Craig Le Houllier of Raleigh NC, from seed sent to him by D. Green of Servierville, Tennessee, whose ancestors received the strain from the local Cherokee people over 100 years ago.

This variety crops well even in hot dry conditions with the 11oz/300gm slightly flattened fruits ripening to a unique deep, dusky, pink-purple colour with greenish shoulders. The colour goes right through the flesh which has a sweet, deep, rich, complex smoky quality all of its own.

Nyagous

Type Productive cordon   80 – 85
Site Cold greenhouse or very sunny spot outdoors.
Fruit Medium sized, dark brownish red.
Plus points Good crops of sweet-sour smoky flavoured tomatoes.

Originally from Australia, Nyagous is an unusual 'Black' tomato; although it is included among them it is not really black, more a dark brownish brick red with an emerald green inside. One of the attractions of Nyagous is its skin which is smooth and unlike many black tomatoes is not prone to splitting or cracking and sometimes a greyish tint is seen across the sides of the fruit.

The flavour of Nyagous is sweet and slightly sour with enticing smoky under-tones. Like other varieties of similar colour, there is little indication of ripeness, so feel for a slight softening of the shoulders and pick before it is too ripe.

Purple Calabash

Type Very vigorous cordon.   80 – 90
Site Cold greenhouse or very warm spot outdoors.
Fruit One of the strangest looking tomatoes, with a complex fascinating flavour.
Plus points Heavy crops of intriguing fruit, a connoisseur's choice.

During the first season that I grew Purple Calabash, the thin skinned fruit had 2 different flavour patterns and it took another full season's weather to identify the cause: prolonged hot sunshine. The flavour of most varieties of tomato improves in sunny weather but when Purple Calabash goes through the mature green/ ripening/ fully ripe stages in hot sunshine it develops the most amazingly rich, fruity, sweet, with a hint of complex spice, truly mouthwatering flavour.

The spiciness decreases during mixed weather and after a prolonged spell of dull weather that glorious complexity disappears from the green gel surrounding the seeds. The plant is a heavy continuous cropper and when the sun returns so does the flavour to these deeply fluted, puckered, flattish fruits.

WHITE TOMATOES

Grosse Blanche

Type Cordon.
Site Hot or cold greenhouse.
Fruit Large white tending towards cream.
Plus points Sweet juicy, mild flavour.

Whereas most gardeners seek strongly flavoured tomato varieties with a good sugar-acid balance, not everyone enjoys such robust flavours. Fortunately there are varieties which offer a sweet succulent fruit for people with a more delicate palate, or who can appreciate the subtle nuances of these intriguing white tomatoes.

Tomato colour is a mixture of the pigments in the skin and the colour of the flesh beneath. When a yellow tomato is covered by a colourless skin the colour of the fruit is determined by the colour of the flesh beneath. Around 20 varieties of white tomatoes are currently available, varying from transparent white through ivory to yellowy white and from cherries to beefsteak in size.

Grosse Blanche is one of several large fruited white tomatoes typical of the type: it is moderately vigorous, with each truss carrying 2 to 4, 3-4 "/8-10cm sweet, juicy, mild flavourful white tomatoes. When is a white tomato ripe? They vary, but when the flesh round the shoulder feels soft, try them, some are better picked just under ripe, others when fully mature. It is a case of trial and error until the best stage for each variety is found.

Ivory Egg

Type Vigorous, strong cordon.
Site Hot or cold glasshouse.
Fruit True egg shape and colour.
Plus points Unusual shape, solid mild flesh.

This vigorous cordon well suited to growing in cold houses originally came from the USA. Ivory Egg is named as each fruit is the size and shape of an egg; solid and meaty in texture, the flesh is low in acid and appreciated by those who enjoy a more mild yet flavoursome tomato.

White as it starts to ripen, Ivory Egg turns a yellowish ivory when ripe, the fruit are better picked when just coming up to ripeness as when fully ripe are easy to knock off the truss when passing.

Green Tomatoes

Aunt Ruby's German Green

Type Vigorous cordon   80 – 85
Site Inside or outdoors.
Fruit Beautiful green beefsteak.
Plus points Vigorous and prolific with sweet juicy flesh.

Aunt Ruby Arnold of Greenville Tennessee lived to be 92; whether her longevity was due to eating German Green tomatoes is not recorded. What is known however is that Aunt Ruby got this variety from her grandfather who brought it from Germany.

German Green is a beautiful fruit tipping the scales at around 12oz/400 gm, the slightly flattened fruits ripen to a pale green with yellow sides and a pink blush that extends into the flesh. Aunt Ruby's strain has a chartreuse golden heart with the seeds coated in lime green gel.

After admiring this beauty, the sweet juicy flesh has a refreshing spicy flavour. An elixir? well, Aunt Ruby did live to be 92.

Green Bell Pepper

Type Vigorous cordon   75
Site Cold greenhouse or outdoors.
Fruit Size and shape of a bell pepper.
Plus points Grown for stuffing.

Green Bell pepper is well named as it resembles a pepper, not only outside but inside too where it is hollow. This most attractive fruit has dark green mottled stripings which turn yellow green when ready to eat, when but when over ripe turn yellow.

Originally from the USA, the plant produces lots of 'bells' which are grown solely for the kitchen, where, filled with any savoury mixture and baked they are delicious.

Green Grape

Type Loose bush   70 – 75
Site Cold greenhouse or outdoors.
Fruit   Sweet,   green,   juicy
1"/2.5cm 'grapes'.'
Plus points The only green cherry, delicious, does well in pots.

It has to be said that Green Grape is a somewhat loose, wayward grower of a bush and although it can be left to sprawl, the 2-3 foot/60-90cm plant is better staked and tied with a loop of twine, or better still planted in a pot.

Although the ancestors of Green Grape are heirloom varieties, this particular cultivar is only 20 years old, having been introduced by Messrs. Tater Mater in 1986. By mid season the untidy plant is covered with bunches of 6 to 12 most unusual grape shaped fruit. When ripe, the shoulders of each 'grape' are green turning to green with yellow veins, becoming a translucent pale green at the bottom. They are full of luscious spicy sweet juice and are readily picked and eaten straight from the plant.

Green Zebra

Type Cordon   75
Site Cold greenhouse or outdoors.
Fruit Well named, delectable.
Plus points Medium vigour, prolific, some disease resistance.

Green Zebra was the unlikely star of our 2005 tastings, when people found its appearance fascinating and its flavour intriguing and wanted more. What a difference a decade makes; when I first grew Green Zebra some 10 years ago an offer to taste was usually met with 'Green tomatoes?', 'No thanks', nowadays it is more likely to be 'Mm can I try another piece?'

This is another variety bred from several heirloom ancestors in 1985 by Tom Wagner of Tater Mater Seeds in the USA. The plant is moderately vigorous, carrying 4 to 6 tomatoes on well spaced trusses, the 2"/5cm fruit have pronounced dark green zebra like stripes. When ripe the shoulders of Green Zebra turn a golden yellow between the dark green stripes and when cut. the inside is lime/emerald green with bright emerald gel. The taste is rich, tangy, and tart with a sweet spicy lemon-lime complexity. Served either in wedges or sliced across it certainly spices up a salad.

Tasty Evergreen

Type Cordon   80
Site Cold glasshouse or outdoors.
Fruit Medium large, light yellowish green.
Plus points Delicious, mild intriguing flavour.

Tasty Evergreen is one of the great heirloom tomatoes collected by the late Ben Quisenbury in the USA. and it poses the question, 'When is Tasty Evergreen ripe?'.

There comes a time during the ripening process when the shoulders of the medium to large fruits start to turn to a paler shade of light yellowish green; πpick one and try it. The ripe fruit when cut has a glorious emerald green flesh of a mild yet distinctive, delicate flavour. Not a very heavy cropper, but a delightfully different taste.

List of suppliers

All the varieties of tomatoes listed are available as seeds or plants from one or several of the suppliers below.

D.T. Brown & Co. Ltd. Bury Rd, Newmarket. CB8 7PQ 0845 1662275

S. Dobie & Sons. Long Rd, Paignton, Devon. TQ4 78X 0870 1123625

Chilton Seeds. Bortree Stile, Ulverstone. Cumbria. LA12 7PB 01229 581137

Thomas Etty Esq. 45 Forde Av., Bromley, Kent. BR1 3EU 020 8466 6785

Mr. Fothergills Seeds. Kentford, Suffolk. CB8 7QB 0845 1662511

HDRA Heritage Seed Library. Ryton Gardens, Coventry. CV8 3LG 024 7630 3517

Innovar Plant Breeding. Station Rd, Reepham, Norfolk. NR10 4NB 01603 870541

S.E.Marshall & Co. Alconbury Hill, Huntingdon, Cambs. PE28 4HY 01480 443390

Organic Gardening Catalogue. Hersham, Surrey. KT12 4RG 0845 1301304

Real Seed Catalogue. Brithdir Mawr, Newport, Pembs. SA42 0QJ 01239 821107

Robinsons. Forton, Preston, Lancs. PR3 0BR 01524 791210

Seeds of Italy. Phoenix Ind. Est., Rosslyn Cres., Harrow. HA1 2SP 020 8427 5020

Select Seeds. 58 Bentinck Rd, Shuttlewood, Chesterfield. S44 6RQ 01246 826011

Simpsons Seeds. Horningsham Warminster, Wiltshire BA12 7NQ   01985 845004

Suffolk Herbs. Monks Farm, Kelvedon, Essex. CO5 9PG 01376 572456

Suttons Seeds. Hele Rd, Torquay, Devon. TQ2 7QJ 01803 69636

Tamar Organics Gulworthy, Tavistock, Devon. PL19 8JE 01822 834690

Terre De Semences with Association Kokopelli. Ripple Farm, Crundale, Canterbury, Kent. CT4 7EB 01227 731815

Edwin Tucker Stonepark, Ashburton, Devon. TQ13 7DG 01364 652233

Thompson & Morgan. Poplar Lane, Ipswich, Suffolk IP8 3BU 01473 688821

Unwins Ltd. Wisbech, Cambridgeshire PE13 27X 01945 588522

For information about supplementary and replacement greenhouse lighting systems, details available from

The Farm Energy Centre,   National Agricultural Centre, Stoneleigh Park, Kenilworth. Warwickshire. CV8 2LS   024 7669 6512

Biological Pest Control Suppliers,

Organic Seed Catalogue.   )
Suffolk Herbs.   )   Addresses as above
Tamar Organics.   )
Edwin Tucker & Sons Ltd.)

Many Garden Centres also offer a list of contacts for supplies of a range of Biological Pest Controls.

Index

Tomato varieties entered in bold type